P9-ARP-983

Adobe Photoshop

NIGEL FRENCH
MIKE RANKIN

 Peachpit Press

Adobe Photoshop
Visual QuickStart Guide
Nigel French and Mike Rankin

Peachpit Press
www.peachpit.com

Peachpit Press is an imprint of Pearson Education, Inc.

San Francisco, CA

To report errors, please send a note to errata@peachpit.com

Notice of Rights

Notice of Liability

Trademarks

Executive Editor: Laura Norman
Development Editor: Linda Laflamme
Technical Editor: Conrad Chavez
Senior Production Editor: Tracey Croom
Copy Editor: Linda Laflamme
Compositor: Danielle Foster
Proofreader: Kelly Kordes Anton
Indexer: James Minkin
Cover Design: RHDG / Riezebos Holzbaur Design Group
Peachpit Press
Interior Design: Peachpit Press with Danielle Foster
Logo Design: MINE™ www.minesf.com

ISBN-13: 978-0-13-764083-6
ISBN-10: 0-13-764083-8

1 2022

Dedication

Nigel: To Melanie

Mike: To Shirley Rankin

Special Thanks

The authors offer special thanks to Linda Laflamme for her editing wisdom, patience, and good humor, and Conrad Chavez for his expertise and insightful technical editing. Also, thanks to Laura Norman and Tracey Croom for shepherding this project from its inception through to publication.

Our gratitude also extends to Melanie Hobson and Laura Miller for allowing use of their photos, and to all our colleagues and friends in the creative community. May we all continue to inspire and learn from each other.

Pearson's Commitment to Diversity, Equity, and Inclusion

Pearson is dedicated to creating bias-free content that reflects the diversity of all learners. We embrace the many dimensions of diversity, including but not limited to race, ethnicity, gender, socioeconomic status, ability, age, sexual orientation, and religious or political beliefs.

Education is a powerful force for equity and change in our world. It has the potential to deliver opportunities that improve lives and enable economic mobility. As we work with authors to create content for every product and service, we acknowledge our responsibility to demonstrate inclusivity and incorporate diverse scholarship so that everyone can achieve their potential through learning. As the world's leading learning company, we have a duty to help drive change and live up to our purpose to help more people create a better life for themselves and to create a better world.

Our ambition is to purposefully contribute to a world where:

- Everyone has an equitable and lifelong opportunity to succeed through learning.
- Our educational products and services are inclusive and represent the rich diversity of learners.
- Our educational content accurately reflects the histories and experiences of the learners we serve.
- Our educational content prompts deeper discussions with learners and motivates them to expand their own learning (and worldview).

While we work hard to present unbiased content, we want to hear from you about any concerns or needs with this Pearson product so that we can investigate and address them.

- Please contact us with concerns about any potential bias at https://www.pearson.com/report-bias.html.

Contents at a Glance

Image Credits

Table of Contents

TIP See page xix for information on accessing this **Bonus Content.**

LIST OF VIDEOS

LIST OF VIDEOS

LIST OF VIDEOS

LIST OF VIDEOS

Introduction

Welcome to Adobe Photoshop, the industry standard image-editing software used by artists, graphic designers, photographers, and pretty much anyone who wants to make their images the best images they can be.

Today, Photoshop is so much a part of our cultural landscape that it is often used as a verb. Yet, the wealth and breadth of Photoshop features can be intimidating—not just to new users, but also to seasoned veterans who may not have kept up with the application's continuing evolution.

We hope you'll find this reference a trusty companion that guides you through the essential tasks you need to fulfill with Photoshop.

In This Introduction

How to Use This Book

This Visual QuickStart Guide, like others in the series, is a task-based reference. Each chapter focuses on a specific area of the application and presents it in a series of concise, illustrated steps. We encourage you to follow along using your own images.

This book is meant to be a reference work, and although we don't expect readers to read in sequence from start to finish, we've ordered the chapters in a logical fashion to build on each other.

We start out with the fundamentals of opening and saving documents, navigating your way around a document, and knowing what stuff is called. From there, we move on to some fundamentals and conventions of image editing before delving into common tasks: cropping, making selections, and working with layers, layer masks, and adjustment layers. It's all here: blend modes, color modes, retouching images, using the Adobe Camera Raw plug-in, taking advantage of Smart Objects, exploring filters, working with type, and more.

This book is suitable for beginners as well as intermediate users of Photoshop who'd like a refresher on specific topics.

Enhance and Amplify

This book can be used with and without the videos, but the videos expound on certain things. Videos are available to all who purchase this product, but are accessible only as part of the included Web Edition.

Sharing Space with Windows and macOS

Photoshop is almost exactly the same on Windows as it is under macOS, which is why this book covers both platforms. The biggest visible difference is the different location of the Preferences menu: at the bottom of the Edit menu on Windows, and under the Photoshop application menu on macOS.

The screenshots in the book have been captured with a light interface just because it tends to reproduce better in print.

We frequently mention keyboard short-cuts, which are great time-savers, and these are listed with the Windows version appearing first followed by a forward-slash and then the macOS version. For example, a simple keyboard shortcut appears as Ctrl/Command+C, and a complicated short cut appears as Ctrl+Alt+Z/Command+Option+Z.

VIDEOS
Sample Video Title

Video icon makes it easy find the video to watch in the Web Edition.

Online Content

Your purchase of this Visual QuickStart Guide includes a free online edition of the book, which contains the companion videos and three additional chapters of bonus material on Adobe Camera Raw, Adobe Bridge, and working with layer styles and layer effects. You can access the Web Edition from your Account page on www.peachpit.com.

Web Edition

The Web Edition is an online interactive version of the book, providing an enhanced learning experience. You can access it from any device with a connection to the internet, and it contains the following:

- The complete text of the book
- Hours of instructional video keyed to the text
- Bonus chapters

Accessing the Web Edition and Bonus Content

Note: If you encounter problems registering your product or accessing the Web Edition, go to www.peachpit.com/support for assistance.

You must register your purchase on peachpit.com in order to access the online content:

1. Go to www.peachpit.com/ photoshopvqs2022.

2. Sign in or create a new account.

3. Click Submit.

4. Answer the question as proof of purchase.

5. The Web Edition can be accessed from the Digital Purchases tab on your Account page. Click the Launch link to access the product.

6. Access the **Bonus Content** from the Registered Products tab on your Account page.

If you purchased a digital product directly from peachpit.com, your product will already be registered, but you still need to follow the registration steps to access the Web Edition.

1

Getting Started

When you open Photoshop for the first time, its interface can be rather intimidating: like being in the helicopter cockpit with no idea of how to fly the thing. So, before we take off, let's do some essential background work. In this chapter, you'll get acquainted with some of the fundamentals of working in Photoshop: opening existing files, creating new files from scratch, saving files, and ending a work session. We may not leave the ground, but you'll feel much more confident about doing so when the time comes.

In This Chapter

Opening Files

There are several ways to open files: using the Open command, using the Open Recent command, or dragging a compatible file format onto the Photoshop application icon.

To open a file:

1. Choose File > Open (Ctrl/Command+O), or click the Open button on the Home Screen (**FIGURE 1.1**).

2. Navigate to the folder containing the file and from the file list, select the file you want to open. If the file does not appear, select the option for showing all files from the Files Of Type (Windows) or Enable (macOS) menu.

3. Click Open. In some cases, a dialog appears, letting you set format-specific options.

TIP If you open multiple files, they will appear as tabbed documents. Click a tab to switch between the open documents. You can also choose the name of any open document on the Windows menu.

TIP Sometimes it's easier to choose from a list of recently opened files, either from the File > Open Recent submenu or on the Home Screen. (Note that the Home Screen will be visible only if Auto Show the Home Screen is selected in General Preferences.)

TIP If you choose a file in camera raw format, it will open in Adobe Camera Raw. After you adjust it there, you can open it in Photoshop. (See Chapter 20, online.)

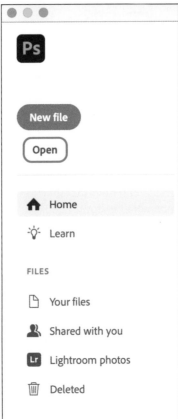

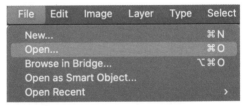

FIGURE 1.1 You can open a file from the File menu (top) or the Home Screen.

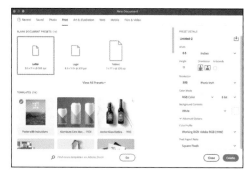

FIGURE 1.2 The New Document dialog

FIGURE 1.3 Choose a tab to display a selection of blank document presets that most closely match the type of document you want to create.

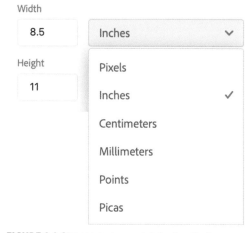

FIGURE 1.4 Choose a size and default unit of measurement.

FIGURE 1.5 If the default values don't fit your needs, choose a new resolution, color mode, and bit depth for your document.

Creating a New, Blank Document

When you create a new document there are a number of decisions to be made, some of which may not make sense if you're new to Photoshop. If in doubt, use the default settings for now.

To create a new document:

1. Choose File > New (Ctrl/Command+N), or click Create New File on the Home Screen to open the New Document dialog (**FIGURE 1.2**).

2. At the top of the dialog, click a document type tab (Photo, Print, Art & Illustration, Web, Mobile, or Film & Video), and then select one of the available document presets that appear (**FIGURE 1.3**).

3. Type a name in the Name field, or you can skip this and name the file when you save it.

4. If the default document size isn't what you need, choose a unit of measurement, and then enter values for Width and Height (**FIGURE 1.4**).

5. Choose a document orientation: portrait (tall) or landscape (wide).

6. Enter the Resolution required for your target output condition (print or screen, for example).

7. Choose a color mode and bit depth; for most projects use RGB and 8-bit respectively (**FIGURE 1.5**).

8. (Optional) Click the Background Contents color swatch to choose a custom background color. White is the default, and there is rarely a need to change this; you can easily change the color of the background once the document is in progress.

9. (Optional) Under Advanced Options, choose a profile from the Color Profile menu (**FIGURE 1.6**). The available profiles will vary depending on the document's Color Mode setting. For print output, we recommend choosing Adobe RGB (1998); for web output, we recommend sRGB. (For print and web output, leave Pixel Aspect Ratio set to Square Pixels.)

10. Click OK. A new, blank document window appears.

TIP If the Clipboard contains image data (artwork that you've copied from Photoshop or Illustrator or from a web browser), the New Document dialog uses the pixel dimensions of the Clipboard image contents as the default Width and Height.

Document Presets

If you tend to choose the same document size, color mode, and other settings, you can save time by creating a document preset.

To create a document preset:

1. Choose File > New (Ctrl/Command+N) to open the New Document dialog.

2. Choose the settings you want.

3. Click the Save Document Preset button.

4. Type a name for your preset. The new preset will now be listed under the Saved tab in the New Document dialog (**FIGURE 1.7**).

TIP To delete a user-created preset, choose it then click the Trashcan icon in the top right of its thumbnail.

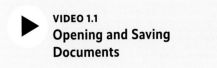

> ▶ **VIDEO 1.1**
> **Opening and Saving Documents**

FIGURE 1.6 The Advanced Options in the New Document dialog.

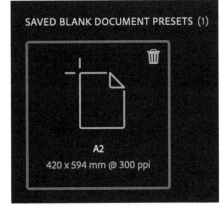

FIGURE 1.7 A saved preset

Overview: New Document Dialog

The New Document dialog lets you:

- Create documents using blank document presets for multiple categories. Before using the presets, you can modify their settings.

- Access recently used templates, and items (click the Recent tab to see a list).

- Save your own custom presets and access them later (click the Saved tab to see them).

- Create documents using templates from Adobe Stock.

Templates

Rather than begin with a blank canvas, you can start with a template from Adobe Stock. Some Adobe Stock templates are free; others are paid. Templates include assets and illustrations that you can build on to complete your project. You can work with a template just as you would any other Photoshop document (PSD).

To use a template:

1. In the New Document dialog, click a category tab: Photo, Print, Art & Illustration, Web, Mobile, or Film & Video.

2. Select a template (**FIGURE 1.8**).

FIGURE 1.8 Some of the free templates available from Adobe Stock

3. Click See Preview to view a preview of the template.

4. Click Download. Photoshop prompts you to license the template. After you do so, you download it.

5. After the template has downloaded, click Open to open it. While the template opens, you may be prompted to sync fonts from Adobe Fonts; click OK.

•   Downloaded templates are added to a Creative Cloud Library called Stock Templates. You can access this library in the Libraries panel. They also appear in the New Document dialog under Recent > Templates.

•   When you open a template, an instance of it is opened as a new Untitled PSD document. Changes made to that PSD document do not affect the original template.

•   You can use the Find More Templates On Adobe Stock field to search for and download additional templates. Photoshop opens the Adobe Stock website in a new browser window.

Downloading Images from a Camera

One way of "getting images into Photoshop" is to download images to your computer by connecting your camera or a media card reader. Once the files are on your computer, you can use your favorite method to open them as Photoshop documents.

To load images from your camera:

Do one of the following:

- Use the Get Photos from Camera command in Adobe Bridge to download photos (**FIGURE 1.9**), as well as organize, rename, and apply metadata to them. (See Chapter 21.)

- If your camera or the card reader appears as a drive on your computer, copy image files to your hard drive.

- Use the software that came with your camera, Windows Image Acquisition (Windows), or Image Capture (macOS) to transfer the images to your computer.

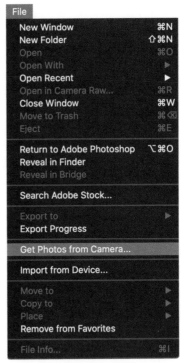

FIGURE 1.9 The Bridge File menu, from which you can download photos

Acquiring Images from a Scanner

Scanners don't get much love these days, but they are still a good way to digitize images. Most scanners come with standalone software to scan and save images. No matter which you use, the basic procedure is the same:

1. Start the scanning software, and set options as desired.

2. Save scanned images in TIFF format.

3. In Photoshop, open the saved TIFF files.

Some scanner software lets you designate Photoshop as the external editor for an image after scanning is completed, as well. Consult your scanner documentation for more details.

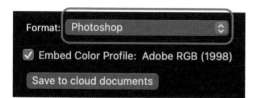

FIGURE 1.10 Choosing a file format

File Saving Options

The options Photoshop offers when you choose File > Save As depend on the image you're saving and the selected file format (**FIGURE 1.11**).

Save a Copy saves a copy of the file while keeping the current file open.

Alpha Channels saves any alpha channel that may have been added to the document. (See Chapter 8.)

Layers preserves the image layers. If you turn off this option or it's unavailable, layers are flattened or merged (depending on the selected format). (See Chapter 6.)

Notes saves any notes created with the Notes tool with the image.

Spot Colors saves any spot channels that may have been added to the document. (See Chapter 11.)

Use Proof Setup, ICC Profile (Windows), or Embed Color Profile (macOS) creates a color-managed document. (See Chapter 19.)

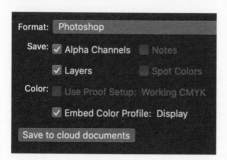

FIGURE 1.11 The file saving options for a native Photoshop (PSD) file

Saving Documents

Although Photoshop lets you save files in more than a dozen formats, you'll probably use just a few of them, such as PSD (the native Photoshop file format), Photoshop PDF, JPEG, and TIFF. If you're not sure what format to use, stick with the PSD format. In fact, to work nondestructively (more on this later) you should always save your document in PSD format, and then save copies in whatever format you require. This way you'll always have an editable "master file" PSD to return to when necessary.

To save a document:

1. Choose File > Save (Ctrl/Command+S). If the document is blank, choose File > Save As (Ctrl+Shift+S/Command+Shift+S). The Save As dialog opens.

2. Type a name in the File Name field (Windows) or the Save As field (macOS).

3. Click the On Your Computer button.

4. Choose a location for the file. In Windows, use the Navigation pane on the left side of the dialog. In macOS, click a drive or folder in the Sidebar panel on the left side of the window, then click a folder in one of the columns. To locate a recently used folder, use the menu below the Save As field.

5. From the Save As Type/Format menu, choose a file format (**FIGURE 1.10**). The PSD, PSB (Large Document Format), Photoshop PDF, and TIFF formats support layers.

6. If you're not yet familiar with the features listed in the Save area and the Color area, leave the settings as they are.

7. Click Save.

After you save a file for the first time, each subsequent use of the Save command overwrites the previous version by default. You can, however, easily save a separate copy if you need multiple variations of an image. The Save a Copy command creates a separate copy while leaving the existing document open.

To save a separate copy of an existing document:

1. Choose File > Save A Copy (Ctrl+Alt+S/ Command+Option+S) to save a copy of a file under a new name or with different options (e.g., with or without alpha channels or layers).

2. Check any available options in the Save area that are appropriate.

3. (Optional) The word *copy* is appended to the filename by default. If you prefer, give the file a distinct name.

4. Click Save.

TIP An asterisk on a document tab or title bar indicates unsaved changes.

TIP To have the location in the Save As dialog default to the location of the current file, go to Edit/Photoshop > Preferences > File Handling, and check Save As To Original Folder.

TIP Another way to copy an entire document is to choose Image > Duplicate. This creates a new window for the copy of the file, which you'll then need to save as a separate file.

To save a flattened copy of a file:

1. Choose File > Save A Copy.

2. Deselect the Layers option; this becomes automatically deselected if you choose a format, like JPEG, that does not support layers.

3. Choose a format and any available options that are appropriate.

4. Click Save.

File Saving Preferences

Using the File Handling preferences, you can customize your default file saving options. Because there are so many, we're just going to cover the most important settings.

Choose Edit/Photoshop > Preferences > File Handling, then under File Saving Options choose any of the following (**FIGURE 1.12**):

FIGURE 1.12 Customize file saving options in File Handling Preferences.

- **Image Previews:** Choose Always Save to include file previews automatically (our preferred option). If you choose Ask When Saving, this preview option will display in the Save As dialog instead.

- **Save In Background:** When this is turned on, you don't have to wait for Photoshop to finish saving the file before you continue working.

- **Automatically Save Recovery Information:** Turn this option on to automatically store crash-recovery information at a specified interval. If you experience a crash, when you restart Photoshop recovers your work.

Cloud Documents

PSDC, Adobe's cloud-native document file type, allows you to work across devices online or offline. With cloud documents your edits are saved and synced to the cloud. So long as you're signed into your Adobe account, you can work on them anywhere, using any device.

On Photoshop on the desktop, for example, you can choose to save your document as a cloud document so that you can edit the document using Photoshop on the iPad (**FIGURE 1.13**). Choose File > Save As, and (if the dialog is not already in the Save to Creative Cloud mode) click the Save To Cloud Documents button.

Photoshop on the iPad saves your documents as PSDC cloud documents by default.

You can also access cloud documents from within Photoshop and on the web at assets.adobe.com.

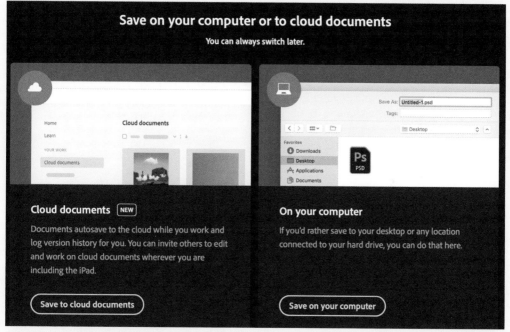

FIGURE 1.13 Choose where you want to save your documents—to the cloud or to your computer. Select Don't Show Again so that you don't need to make this choice every time you save a document. Photoshop remembers your decision and gives you the same option until you choose otherwise.

Closing Documents and Quitting Photoshop

When you're finished working on a document, be sure to close it to make memory available for other documents—and so that you don't inadvertently edit the document. When it's time to close up shop for the day, you can quit Photoshop.

To close a document:

1. Click the X in a document tab, choose File > Close (Ctrl/Command+W), or click the Close button in the corner (upper-right, Windows; upper-left, macOS) of a floating document window.

2. If the file has unsaved changes, an alert appears. Click No (N)/Don't Save to close the file without saving, or click Yes (Y)/Save (S) to save the file (or click Cancel or press Esc to dismiss the Close command).

TIP To close all open documents, choose File > Close All (Ctrl+Alt+W/ Command+Option+W). If an alert dialog appears, you can check Apply To All, if desired, to have one response apply to all the open documents, then click No/Don't Save or Yes/Save.

To exit/quit Photoshop and close all open files:

- In Windows, choose File > Exit (Ctrl+Q), or click the Close button for the application frame.

- In macOS, choose Photoshop > Quit Photoshop (Command+Q).

TIP If any open files contain unsaved changes, an alert dialog appears for each one when you exit/quit Photoshop.

The Status Bar

As you work, keep an eye on the Status bar and menu at the bottom of the document window, which display useful information about the current document.

You can specify what information to display with the menu to the right of the Status bar at the bottom of the Application frame (**FIGURE 1.14**). The most useful pieces of information are:

Document Sizes shows the approximate storage size of a flattened version of the file if it's saved in the PSD format (on the left) and the file size including layers and any alpha channels (on the right).

Document Profile shows the color profile embedded in the current file and the number of bits per channel. If the document doesn't have an embedded profile, it is shown as untagged.

Document Dimensions shows the image dimensions (width and height) and resolution.

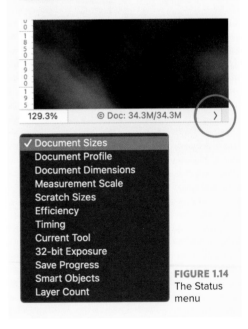

FIGURE 1.14 The Status menu

Documents and Navigation

To be comfortable working in Photoshop you need to know what things are called, how to move around, and how to find the things you need. This should be effortless—and with practice it will be—so that you can concentrate on being creative. In this chapter, you'll get familiar with the Photoshop interface and gain an understanding of why things are named the way they are, as well as why they are located where they are. Crucially, you'll learn how to move around—including how to change the zoom level and screen mode and rotate the canvas view. As you'll see, when it comes to viewing and moving around within documents, there are multiple ways to do things, which is a recurring theme in Photoshop.

The Photoshop Interface

A major component of the Photoshop interface is its many many floating panels that you can rearrange, hide, and show. The good news is that panels are easy to hide or collapse so they don't get in the way when you don't need them, and they are easy to show and expand when you do. You can also organize panels into groups and "dock" them on the side of the screen.

To hide (or show) panels:

Do one of the following:

- Press Tab to hide or show all open panels, including the Tools panel.
- Press Shift+Tab to hide or show open panels *except* the Tools panel.

To find a lost panel:

- Choose the panel's name from the Window menu to show it. The panel displays either in its default group and dock or its last open location.

To minimize and maximize a floating panel group:

1. Double-click the panel bar or tab to minimize the panel group.

2. Double-click the panel bar again to maximize the panel or group, or simply click the panel tab (**FIGURE 2.1**).

> **TIP** When using a Brush tool, you can show the Brush Settings panel by clicking the Toggle Brush Settings Panel button on the Options bar.

> **TIP** When using a Type tool, you can show the Character/Paragraph panel group by clicking its toggle button on the Options bar.

To close a panel:

Do one of the following:

- Right-click the panel tab and choose Close or Close Tab from the context menu.
- Click the Close button in the upper-left of a floating panel.

To group and move panels:

1. Click a panel's tab and drag it to another group.

2. Release when the blue line appears.

3. Drag the panel name left or right to change the panel's position in the group.

FIGURE 2.1 Double-click to minimize/maximize a panel or panel group or to collapse to icons.

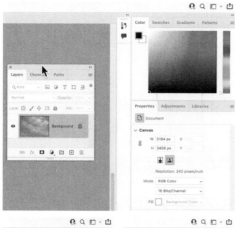

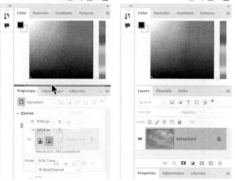

FIGURE 2.2 Docking the Layers panel group: before, during, after

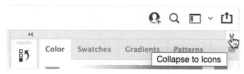

FIGURE 2.3 Collapse to icons

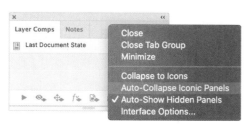

FIGURE 2.4 To access the Auto-Collapse Iconic Panels option, right-click any panel tab, bar, or icon.

To dock and move panels:

1. Drag the panel's tab or bar to the edge of another panel, panel group, or Application Frame.

2. Release when the blue vertical or horizontal (depending on where you want the panel to go) line appears (**FIGURE 2.2**).

3. (Optional) To move a panel group up or down in a dock, drag the panel bar to reposition it, then release when the blue line appears.

To float a docked panel or group:

1. Drag the panel out of the docked or grouped panels.

2. (Optional) To prevent it from docking elsewhere as you move it, hold Ctrl/Command as you drag.

TIP To resize a floating panel, drag an edge or the resize box.

To collapse docked panels to icons:

Do one of the following:

- Click the Collapse To Icons (double arrow) button at the top of a dock or double-click the topmost bar to collapse docked panels to icons (**FIGURE 2.3**). When it's collapsed, you can drag the edge of a dock to show or hide its panel names.

- Turn on the Auto-Collapse Iconic Panels option in Workspace preferences to automatically collapse a panel back to an icon when you click outside it (**FIGURE 2.4**). With this preference off, the panel stays expanded when you leave it.

Viewing Images

Typically, when you open an image in Photoshop, the whole image will be visible. This is called Fit On Screen view. The magnification of the image will depend on its pixel dimensions and the size of your monitor. For example, an image with a high pixel count may initially display at only 16.7% of its actual size, while a file with a lower pixel count may display at 50%.

You can view your document at any percentage between 0.26% and 12800%, but the two view sizes that are most important are Fit On Screen view where you can get an overview of the image to evaluate color, contrast, and composition, and 100% view where you can make decisions about the technical quality of the image. It's important to note that 100% means one document pixel to one screen pixel; it does not mean physical size.

To fit an image to the screen:

Do one of the following:

- Use the shortcut Ctrl/Command+0 (zero).
- Double-click the Hand tool in the toolbox.
- Choose View > Fit On Screen.
- Click Fit Screen on the Options bar when you have the Zoom tool or Hand tool selected (**FIGURE 2.5**).

To view images at 100%:

Do one of the following:

- Use the shortcut Ctrl/Command+1.
- Double-click the Zoom tool in the toolbox.
- Choose View > 100%.
- Click 100% on the Options bar for either the Zoom or Hand tool.

To feel comfortable in Photoshop, it's essential to know how to change the image magnification (zooming in to get bigger, zooming out to get smaller), and how to pan around an image that is enlarged beyond its Fit On Screen size. You can find the current zoom percentage in three locations: on the document tab, in the lower-left corner of the document window, and on the Navigator panel (if shown) (**FIGURE 2.6**).

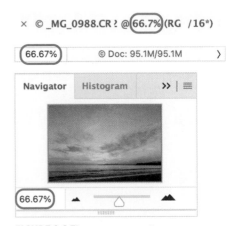

FIGURE 2.6 The zoom percentage on the document tab, the lower left of the document window, and on the Navigator panel

FIGURE 2.5 The Options bar for the Zoom tool

FIGURE 2.7 Scrubby Zoom on the Options bar

FIGURE 2.8
Right-click with
the Zoom tool to
see zoom options.

Zoom Tool Preferences

You can customize how the Zoom tool behaves in Tools preferences. To see your options, choose Edit > Preferences > Tools (Windows) or Photoshop > Preferences > Tools (macOS):

Animated Zoom enables continuous zooming while you hold down the Zoom tool.

Zoom Resizes Window affects the size of the window along with the image when working with floating windows, rather than tabbed documents.

Zoom With Scroll Wheel lets you use the scroll wheel on a mouse to zoom.

Zoom Clicked Point To Center centers the zoom view on the point where you click.

VIDEO 2.1
Working with Panels and Customizing Your Workspace

To use the Zoom tool:

1. Choose the Zoom tool (Z).

2. Click the image to zoom in. Each click enlarges or reduces the image to the next preset percentage, centering the image on the point you click.

3. Alt/Option-click to zoom out.

By default, Animated Zoom is turned on, as well. If you decide you don't like it, you can turn it off in the Tools preferences.

To use Animated Zoom:

1. Click and hold (don't drag) in the image to zoom in.

2. Alt/Option-click and hold to zoom out.

A third zoom option is Scrubby Zoom, which you can turn on from the Options bar (**FIGURE 2.7**). The benefit of Scrubby Zoom is that you don't need to take your hands off the mouse to use it.

To use Scrubby Zoom:

1. Select Scrubby Zoom on the Options bar (if it isn't already on).

2. Click the spot you want to zoom in on, then immediately drag right to zoom in.

3. Click again and immediately drag left to zoom out.

Not everyone loves Scrubby Zoom, however. An alternatively approach is to turn this feature off, then drag over the portion of the image that you want to magnify. The area inside the marquee appears at the highest possible magnification. To move a marquee around an image, begin dragging the marquee, then hold the spacebar.

TIP To adjust the zoom, you can also right-click the image and choose a zoom option from the context menu (**FIGURE 2.8**).

TIP If you reach the maximum magnification (12800%) or minimum size (1 pixel), the magnifying glass appears empty.

TIP Click Fill Screen on the Options bar to have the image fill the window (only part of the image may be visible).

TIP To temporarily zoom an image, hold H to access the Hand tool. Click and hold, then drag the zoom marquee over a different portion of the image and release.

The Navigator Panel

Use the Navigator panel to change the view of your artwork via a thumbnail display. The colored box (proxy view area) represents the currently viewable area in the window.

To use the Navigator panel:

1. Choose Window > Navigator to display the Navigator panel.

2. To change the magnification, type a value in the text box, click the Zoom Out or Zoom In button, or drag the Zoom slider (**FIGURE 2.9**).

3. To change the view of an image, drag the proxy in the image thumbnail.

TIP To simultaneously set the size and position of the proxy area, Ctrl/Command-drag in the image thumbnail.

The Hand Tool

You can use the Hand tool to reposition a magnified image within the document window. For this to work, however, Use Graphics Processor must be turned on in preferences (choose Edit/Photoshop > Preferences > Performance) before you open your document, and Enable Flick Panning must be turned on in Preferences > Tools.

FIGURE 2.9 The Navigator panel: Viewing the whole image (top) and zooming in

FIGURE 2.10 Rotating the view of an image

VIDEO 2.2
Viewing Images

TIP When you choose the Rotate View tool, you can also change the rotation angle of the canvas using the Options bar. Choose the Rotate View tool, then use the scrubby slider, enter a value, or move the dial on the Options bar.

TIP The canvas view can also be rotated using a multi-touch gesture on a track-pad if Enable Gestures is turned on in Preferences > Tools.

To reposition an image with the Hand tool:

1. Press H or hold the spacebar to temporarily invoke the Hand tool.

2. Drag in the document window to move the image.

TIP Select Overscroll in Preferences > Tools to let you scroll past the canvas edge. This is helpful when working with layers that meet or extend beyond the canvas edge.

TIP Select Enable Flick Panning to move an image in the document window by "flick panning" (aka tossing the document around) with the Hand tool.

TIP If you have multiple documents open, you can scroll them simultaneously with the Hand tool by turning on Scroll All Windows on the Options bar.

Rotate View Tool

The Rotate View tool tilts the canvas temporarily so you can work at a more comfortable angle—useful if you are drawing or painting with a stylus on a tablet. The Rotate View tool does not transform the image, only the view of it. To use the Rotate View tool, Use Graphics Processor must be checked in Edit/Photoshop > Preferences > Performance. If the preference is off, check it, then reopen the document.

To rotate the canvas view:

1. Choose the Rotate View tool (in the Hand tool menu) or hold R to spring-load the tool.

2. Drag in the image (a compass displays temporarily) to rotate. Hold the Shift key if you want to constrain the rotation to 15° increments (**FIGURE 2.10**).

3. Click Reset View on the Options bar to reset the canvas.

Tools

The Tools panel—aka the toolbar—is on the left side of the screen. (If it's hidden, choose Window > Tools to display it.) Most of the tools found here have options that appear in the Options bar. To select a tool, simply click it or press its single key shortcut.

You also can expand most tools to show hidden tools that share the same tool space—indicated by a small triangle in the lower right of the tool's icon.

Icons only explain so much, however. To learn more about a tool, position the cursor over the tool to view a tooltip. Some tools offer rich tooltips that display a description and a short video of the tool in action (**FIGURE 2.11**). Once you're more familiar with the tools, you can turn off these rich tooltips by deselecting the Preferences > Tools > Show Rich Tooltips preference.

To access hidden tools in the Tools panel:

Do one of the following:

- Click and hold the visible tool to show hidden tools (**FIGURE 2.12**).

- Alt/Option-click the visible tool to cycle through related tools in the same slot.

- Press Shift plus the single key shortcut to cycle through tools in the same slot (e.g., press Shift+L to cycle through the three Lasso tools). For this to work the Shift Key For Tool Switch option must be turned on in Edit/Photoshop > Preferences > Tools.

- To temporarily access (spring-load) a tool while another tool is selected, long press and hold its single key shortcut. When you release, you'll go back to the previous tool you were working with.

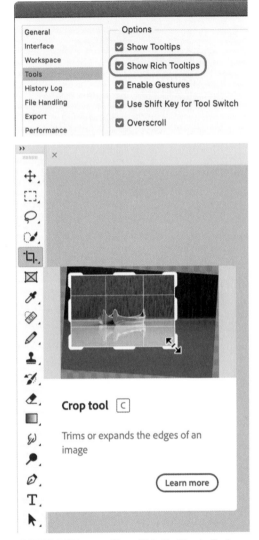

FIGURE 2.11 Turn on Show Rich Tooltips in Tools preferences to see a quick visual explanation of what the tool does.

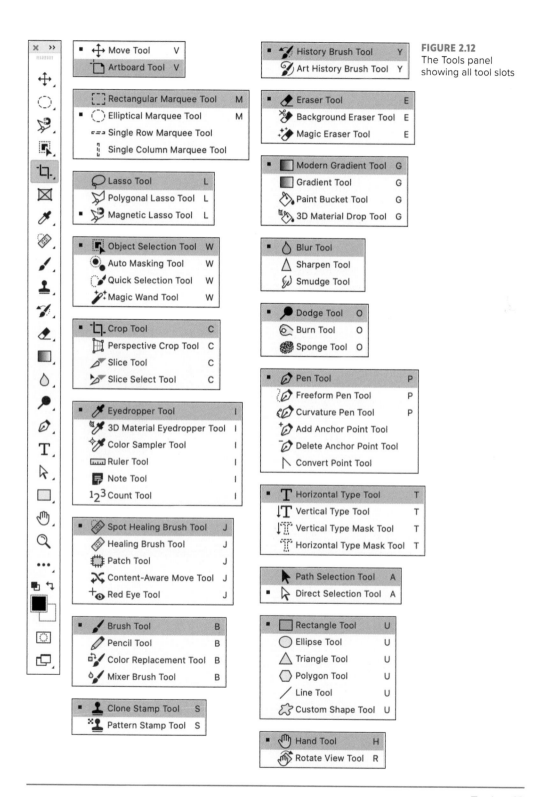

FIGURE 2.12
The Tools panel
showing all tool slots

Tool	Shortcut
Move Tool	V
Artboard Tool	V

Tool	Shortcut
Rectangular Marquee Tool	M
Elliptical Marquee Tool	M
Single Row Marquee Tool	
Single Column Marquee Tool	

Tool	Shortcut
Lasso Tool	L
Polygonal Lasso Tool	L
Magnetic Lasso Tool	L

Tool	Shortcut
Object Selection Tool	W
Auto Masking Tool	W
Quick Selection Tool	W
Magic Wand Tool	W

Tool	Shortcut
Crop Tool	C
Perspective Crop Tool	C
Slice Tool	C
Slice Select Tool	C

Tool	Shortcut
Eyedropper Tool	I
3D Material Eyedropper Tool	I
Color Sampler Tool	I
Ruler Tool	I
Note Tool	I
1 2 3 Count Tool	I

Tool	Shortcut
Spot Healing Brush Tool	J
Healing Brush Tool	J
Patch Tool	J
Content-Aware Move Tool	J
Red Eye Tool	J

Tool	Shortcut
Brush Tool	B
Pencil Tool	B
Color Replacement Tool	B
Mixer Brush Tool	B

Tool	Shortcut
Clone Stamp Tool	S
Pattern Stamp Tool	S

Tool	Shortcut
History Brush Tool	Y
Art History Brush Tool	Y

Tool	Shortcut
Eraser Tool	E
Background Eraser Tool	E
Magic Eraser Tool	E

Tool	Shortcut
Modern Gradient Tool	G
Gradient Tool	G
Paint Bucket Tool	G
3D Material Drop Tool	G

Tool	Shortcut
Blur Tool	
Sharpen Tool	
Smudge Tool	

Tool	Shortcut
Dodge Tool	O
Burn Tool	O
Sponge Tool	O

Tool	Shortcut
Pen Tool	P
Freeform Pen Tool	P
Curvature Pen Tool	P
Add Anchor Point Tool	
Delete Anchor Point Tool	
Convert Point Tool	

Tool	Shortcut
Horizontal Type Tool	T
Vertical Type Tool	T
Vertical Type Mask Tool	T
Horizontal Type Mask Tool	T

Tool	Shortcut
Path Selection Tool	A
Direct Selection Tool	A

Tool	Shortcut
Rectangle Tool	U
Ellipse Tool	U
Triangle Tool	U
Polygon Tool	U
Line Tool	U
Custom Shape Tool	U

Tool	Shortcut
Hand Tool	H
Rotate View Tool	R

TIP To see a tool hint, which is like an expanded tooltip, for the current tool, look at the bottom of the Info panel (**FIGURE 2.13**). If you don't see the tool hint, choose Panel Options from the Info panel menu (upper-right corner of the panel), then select Show Tool Hints.

The Options Bar

On the Options bar, you can choose settings for the current tool. The appearance of the Options bar changes depending on what tool you select. For example, for the Brush tool, you can choose a brush preset, as well as size, hardness, blending mode, and opacity settings. The settings remain in effect until you change them, reset that tool, or reset all tools.

To display the Options bar:

- The Options bar is displayed by default, but if it becomes hidden, choose Window > Options to show it.

To restore tool default settings:

1. Click the Tool Preset Picker (left side of the Options bar).

2. Click the gear icon, then choose Reset Tool (to reset only the current tool) or Reset All Tools (to restore defaults for all tools) (**FIGURE 2.14**).

Cursors

You can choose whether the pointer displays as a crosshair, as the icon of the current tool, or, for some tools, as a circle either the size or half the size of the current brush diameter, with or without a crosshair inside it.

To change the cursor's appearance:

- Choose Edit/Photoshop > Preferences > Cursors (**FIGURE 2.15**).

FIGURE 2.13 The Info panel showing a tool hint for the Magnetic Lasso tool

FIGURE 2.14 Resetting tools. A: Click the down arrow to the right of the tool to open the Tool Preset Picker. B: Click the gear icon to access the menu, which includes options to reset the specific tool or all tools (C).

FIGURE 2.15 Changing the appearance of cursors in Preferences

Customize the Toolbar

You can customize the toolbar to fit your workflow by clicking the three dots at the bottom of the toolbar and selecting Edit Toolbar (**FIGURE 2.16**).

In the Customize Toolbar dialog you can:

- Reorganize the toolbar.
- Move infrequently used tools to Extra Tools.
- Click Save Preset to save a custom toolbar.
- Click Load Preset to open a previously saved custom toolbar.
- Click Restore Defaults to reset the default toolbar.
- Click Clear Tools to move all the tools to Extra Tools.
- Select the widgets to show/hide by clicking them at the bottom of the toolbar.

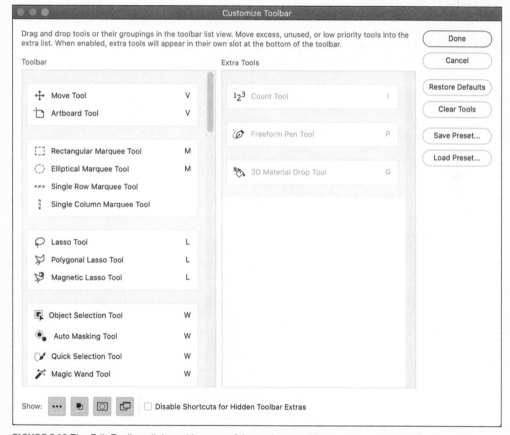

FIGURE 2.16 The Edit Toolbar dialog with some of the tools we seldom use moved to the Extras column

The Application Frame

In Windows, the Photoshop interface is housed in an Application window. It contains a menu bar and Options bar along the top, panels, and any open documents, which by default are docked as tabs. The Application Frame in macOS serves the same purpose but is optional. If the frame is hidden, choose Window > Application Frame. We will refer to both versions generically as the *Application Frame*.

View images as tabbed documents

We recommend docking open document windows as tabs (the default setting) rather than floating them as separate windows. Docked as tabs, you can keep the documents you're not working on out of view but readily accessible (**FIGURE 2.17**). Click the tab to show a docked document.

If documents aren't docking as tabs automatically, you can reset the preference so that they do. Choose Edit/Photoshop > Preferences > Workspace and check Open Documents As Tabs (**FIGURE 2.18**).

To dock a floating document:

1. Drag the document's title bar to the tab area (just below the Options bar) of the Application Frame.

2. Release when the blue line appears.

To dock all floating documents as tabs:

- Choose Window > Arrange > Consolidate All To Tabs. Or, if you have at least one document docked, right-click its tab and choose Consolidate All To Here from the context menu.

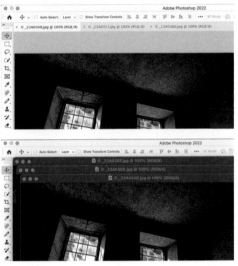

FIGURE 2.17 Three open documents tabbed and the same documents floating in windows

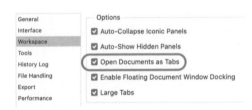

FIGURE 2.18 Workspace preferences

TIP You can resize or minimize the Application Frame. To resize it, drag an edge or corner. To minimize it and any tabbed documents it contains, click the Minimize button (upper right, Windows; upper left, macOS).

TIP Press Ctrl/Control+Tab or Ctrl/Command+~ (tilde) to cycle through open documents. Add the Shift key and you will go in the opposite direction (useful when you have several tabs open and don't want to cycle all the way back to the beginning).

Arrange multiple documents

To view or edit multiple documents simultaneously in the Application Frame, you can arrange them in a preset layout, such as two documents side by side or in a vertical format, or four or six documents in a grid like tiles.

To arrange multiple documents:

Do one of the following:

- From the Window > Arrange submenu, choose a tiling command, such as 2-Up Tile All Vertically, or 4-Up (**FIGURE 2.19**).

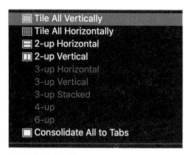

FIGURE 2.19 Choosing a tiling option

- Drag either the bar between two windows or the lower-right corner to resize a window.

- Drag the tab of one document to the tab of another document to reduce the number of visible windows by one.

- Choose Window > Arrange > Consolidate All To Tabs to view one document at a time.

View Image in Multiple Windows

It's possible to have the same image open in more than one window.

To open a new window for a document:

- Choose Window > Arrange > New Window For *[Image File Name]*. You can set the view sizes for each window independently, which is useful when you want to work on a zoomed-in portion of the image while at the same time seeing any changes you make in the context of the whole image (**FIGURE 2.20**).

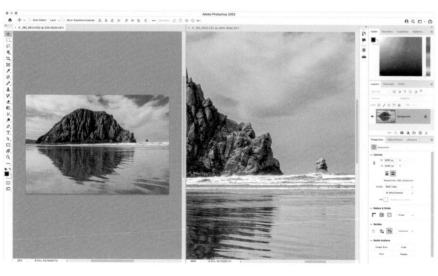

FIGURE 2.20 Viewing an image at different view sizes in two windows simultaneously

To arrange multiple windows:

1. Choose Window > Arrange.
2. Select a display option:
   - ▸ **Cascade** displays undocked windows stacked and cascading from the upper left to the lower right of the screen.
   - ▸ **Tile** displays windows edge to edge. Closing images causes the open windows to resize, filling the available space.
   - ▸ **Float In Window** allows an image to float freely.
   - ▸ **Float All In Windows** floats all images.
   - ▸ **Consolidate All To Tabs** shows the active image in Fit In Window view and minimizes the others to tabs.

Match Zoom and Match Location

The Match Zoom and Match Location options are useful when comparing images, knowing that you're comparing like for like: the same view size and the same part of the image.

To zoom multiple images by the same amount:

1. Open several images, or open one image in multiple windows.
2. Choose Window > Arrange > Tile.
3. With the Zoom tool selected, select Zoom All Windows in the Options bar.
4. Click one of the images. The other images are zoomed in or out by the same relative amount.

Screen Modes

The screen modes control which interface features are shown. To cycle through the screen modes, press F or choose a mode from the Screen Mode menu at the bottom of the Tools panel (**FIGURE 2.21**). The menu offers three choices.

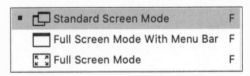

FIGURE 2.21 The Photoshop screen modes

Standard Screen Mode (default) displays the whole interface, including the Application Frame, menu bar, Options bar, current document, document tabs, and panels—with the desktop (and any other open application windows) visible outside it.

Full Screen Mode With Menu Bar displays just the current document with all the above-mentioned interface features showing, but with the Application Frame, document tabs, desktop, and other windows hidden.

Full Screen Mode displays just the current document with the interface features hidden and the panels visible only upon rollover. If shown, the rulers remain visible in Full Screen mode.

FIGURE 2.22 The images compared using the 2-Up Horizontal layout

FIGURE 2.23 Matching the Zoom and Location, we determine that the detail is sharper in the image at right.

FIGURE 2.24 Check Scroll All Windows so that the Hand tool pans all documents simultaneously.

To compare images using Match Zoom:

1. Open two or more images (or open one image in multiple windows). In this example, we're comparing two similar images to determine which to use.

2. Choose Window > Arrange > 2-Up Vertical or 2-Up Horizontal (**FIGURE 2.22**).

3. Choose the Zoom tool and Shift-click one of the images—the other image(s) zooms at the same magnification (**FIGURE 2.23**).

To match location:

1. If the locations of open images are currently different, choose Window > Arrange > Match Location.

2. Choose the Hand tool, select Scroll All Windows in the Options bar, and drag over one of the images (**FIGURE 2.24**). (Alternatively, hold Shift while dragging with the Hand tool.)

Smart Guides

Smart Guides are temporary guides that automatically appear and disappear as you drag a layer or selection. You do not create guides—they become visible if there is a checkmark next to their menu item: View > Show > Smart Guides. With Smart Guides you can:

- Show the distance between layers.

- Show distances from the canvas edge as well as the distances between objects by holding Ctrl/Command while hovering over the shape (**FIGURE 2.25**).

- Align to the top, bottom, left, or right edges, or to the center of objects.

FIGURE 2.25 Using Smart Guides to ensure equal spacing between elements

Precise Positioning

Photoshop has numerous view options to help you position images or elements precisely. Depending on your working style, there will be several that you come to depend upon and use all the time, others that you use less frequently (but are really handy when you need them), and a few that will be "once in a blue moon" options.

Rulers

When visible, rulers appear along the top and left side of the active window in your chosen unit of measurement. Markers in the ruler display the pointer's position.

To show or hide rulers:

- Choose View > Rulers (Ctrl/ Command+R).

Changing the ruler origin or zero point lets you measure from a specific point of an image. The ruler origin also sets the grid's point of origin.

To set the ruler origin:

Do one of the following:

- Start from the upper-left corner where the two rulers meet and drag diagonally into the image (**FIGURE 2.26**).

- Choose View > Snap To, enable any of the commands on that submenu to indicate where you want the ruler origin to snap, then drag the ruler origin.

- Shift-drag diagonally from the upper-left corner into the image to snap the ruler origin to the ruler marks.

To restore the default origin:

- Double-click the upper-left corner.

To change the ruler unit of measurement:

Do one of the following:

- Double-click a ruler, or choose Edit/ Photoshop > Preferences > Units & Rulers, then choose the new unit and click OK.

- Right-click anywhere on the rulers and choose a new unit from the context menu (**FIGURE 2.27**).

- Change the units in the Cursor Coordinates area on the Info panel to automatically change the units on both rulers.

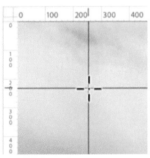

FIGURE 2.26
Changing the origin, or zero point, of the rulers

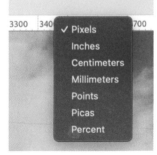

FIGURE 2.27
Right-click the horizontal or vertical ruler to change the unit of measurement.

Guide Behavior

Guides and grids behave in similar ways:

- Selections, selection borders, and tools snap to a guide or the grid when dragged within 8 screen pixels. Guides also snap to the grid when moved.

- Guide visibility and snapping settings are specific to an image.

- Settings for grid spacing, as well as guide and grid color and style, apply to all images.

- If View > Snap To is on, you can snap a guide to a selection, to the edge of content on the currently selected layer, or to a grid line. Having Snap To on is usually preferable, but there may be times when it is restricting. To toggle Snap To off (assuming it is currently on), hold Ctrl while dragging the guide.

- The cursor changes to a double-headed arrow when you drag a guide.

Common guide options are listed on the Properties panel when no layer or artboard is selected (**FIGURE 2.28**).

FIGURE 2.28 Guide options on the Properties panel

Guides and Grid

To help you align images and more, Photoshop offers a grid overlay and several types of guides, which come in different colors and serve different purposes. All guides are non-printing.

Guides (View > Show > Guides or Ctrl/Command+;) show as lines over the image. You can move and remove guides. You can lock them so that you don't move them by mistake.

If working with artboards, there is a distinction between canvas guides (cyan), which cover the whole canvas, and artboard guides (light blue), which show up only on the artboard on which they were drawn.

If you want to add canvas guides to a document containing artboards, click below the artboards on the Layers panel to make sure none are selected, before adding the guide.

Smart Guides (View > Show > Smart Guides) are temporary, magnetic magenta guides that appear when one layer nears the edge of another and can help to align shapes, slices, and selections.

The **Grid** (View > Show > Grid or Ctrl/Cmd-') appears as gray lines or dots and can be useful when adjusting the spacing of multiple elements. The grid is positioned relative to the ruler origin.

To create guides:

Do one of the following:

- Drag a horizontal guide from the horizontal ruler. (Alt/Option-drag from this ruler to create a vertical guide.)

- Drag a vertical guide from the vertical ruler. (Alt/Option-drag from this ruler to create a horizontal guide.)

- Position a guide numerically by choosing View > New Guide, selecting an orientation, and entering a position.

- Choose View > Lock/Unlock Guides to lock and unlock guides.

To create guides from the bounding box of a shape:

- Choose View > New Guides From Shape. This also works with image layers and type layers (**FIGURE 2.29**).

To snap a guide to a target:

Do one of the following:

- Choose View > Snap To > Guides, Grid, Layers, Slices, Document Bounds, or All (of the above).

- With the Move tool (V), drag a selection or layer near a guide, the edge of layer content, or the edge of the canvas. It "snaps" to your target.

To move a guide:

Do one of the following:

- Select the Move tool or hold Ctrl/Command to activate the Move tool.

- Position the pointer over the guide and drag the guide to move it. As you move a guide, its current X or Y position is displayed next to the pointer. (If you can't select a guide, the guides are probably locked. Choose View > Lock Guides.)

FIGURE 2.29 Putting guides around a type layer

- Hold down Alt/Option and double-click a guide to covert it from horizontal to vertical, or vice-versa.

- Hold Shift as you drag a guide to snap to the ruler ticks. If View > Snap To > Grid is selected, the guide snaps to the grid.

To delete guides:

Do one of the following:

- Drag a single guide outside the document window to delete it.

- Choose View > Clear Guides to delete all guides.

To set guide and grid preferences:

1. Choose Edit/Photoshop > Preferences > Guides, Grid, & Slices.

2. Choose a color and style (solid or dashed) for the canvas guides, Smart Guides, and the grid.

3. Enter a value for the grid spacing and subdivisions. Optionally, change the measurement units. You can also choose Percent to create a grid divided into even sections. For example, to divide your canvas into 4 quadrants, set Gridline Every to 100 Percent and the subdivision to 2. Or for a "rule of thirds" overlay, set Gridline Every to 100 Percent and the subdivision to 3 (**FIGURE 2.30**).

FIGURE 2.30 A rule of thirds grid that can be toggled on and off

You can use the New Guide Layout dialog to quickly create modular guide layouts.

To create multicolumn guide layouts:

1. Choose View > New Guide Layout to open the dialog (**FIGURE 2.31**).

2. Check Preview, then enter the number of rows and columns you want.

3. Change the gutter values to specify the space between columns or rows.

4. Check Margin, then enter values in any of the Margin fields to create guides indented in from the top, left, bottom, or right edge of the canvas.

5. Check Clear Existing Guides to delete any previously created guides, including ruler guides (**FIGURE 2.32**).

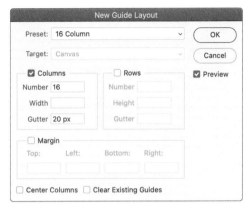

FIGURE 2.31 The New Guide Layout dialog

TIP You can use the scrubby slider over any of the field names to increase or reduce the current value. Hold Alt/Option when using the scrubby slider to change the value in smaller increments, or hold Shift to change the value in larger increments.

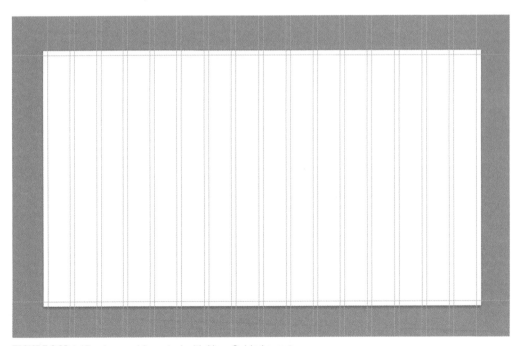

FIGURE 2.32 A 16-column grid created with New Guide Layout

The Ruler Tool

The Ruler tool enables you to calculate the distance between two points. When you measure from one point to another, Photoshop draws a nonprinting line and the Options bar and Info panel show the following information:

- The starting location (X and Y)
- The horizontal (W) distances from the X axis and the vertical (H) distance from the Y axis
- The angle measured relative to the axis (A)
- The total length traveled (L1)
- The lengths traveled (L1 and L2), when you use a protractor

To measure between two points:

1. Select the Ruler tool. (Click and hold the pointer on the Eyedropper tool if the Ruler isn't visible.)

2. Drag from one point to another. Hold Shift to constrain the tool to 45° increments (**FIGURE 2.33**).

TIP Alt/Option-drag at an angle from one end of the measuring line to create a protractor from an existing measuring line. Hold Shift to constrain the tool to 45° increments.

To edit a measuring line:

1. With the Ruler tool, drag one end of an existing measuring line.

2. To move the line, drag from either endpoint.

3. To remove the line, drag the line off the image or click Clear in the Options bar.

The Pixel Grid

At magnifications of 800% or more, an image's pixel grid becomes visible. This can be helpful for creating screen assets when you need to position objects so that they align on a full pixel rather than an anti-aliased edge. To hide the grid, choose View > Show and deselect Pixel Grid (**FIGURE 2.34**).

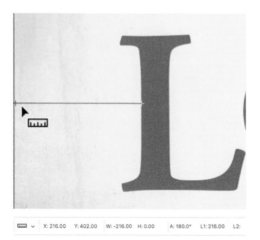

X: 216.00   Y: 402.00   W: -216.00   H: 0.00   A: 180.0°   L1: 216.00   L2:

FIGURE 2.33 Measuring the distance between points

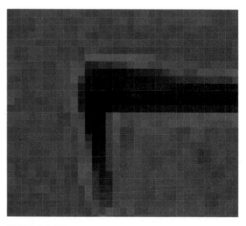

FIGURE 2.34 Viewing the pixel grid at enlarged view sizes

Workspaces

A workspace determines which panels are shown and how they are arranged. To change your panel setup, choose a predefined workspace or a user-defined workspace from the Workspace menu or from the Window > Workspace submenu (**FIGURE 2.35**).

You can also save a custom workspace when you find a panel arrangement that works for your needs. You can also include custom keyboard shortcuts and/or menu sets (menu sets allow you to change the color label and visibility of menu commands). All open panels display in the same location when you relaunch the saved workspace.

To save a custom workspace:

1. Open and position the desired panels in groups and docks the way you want. Close any panels that you rarely use.

2. From the Workspace menu on the Options bar, choose New Workspace.

3. Enter a name for the workspace. Under Capture, turn on Keyboard Shortcuts, Menus or Toolbar, if you customized those elements (**FIGURE 2.36**).

4. Click Save. Your workspace appears at the top of the Workspace menu on the Options bar and on the Window > Workspace submenu.

To edit a workspace:

1. Make the changes desired to your workspace.

2. From the Workspace menu on the Options bar, choose New Workspace.

3. In the New Workspace dialog, retype the same name for your workspace.

4. Click Save, then click Yes.

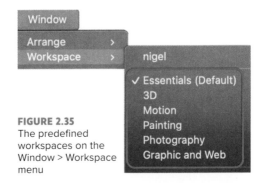

FIGURE 2.35 The predefined workspaces on the Window > Workspace menu

FIGURE 2.36 Create a custom workspace, optionally including keyboard shortcuts, menus, and toolbar modifications.

To reset a workspace:

Do one of the following:

- Choose Reset [workspace name] from the Workspace menu on the Options bar to reset a workspace to how it was.

- Choose Edit/Photoshop > Preferences > Workspace, then click Restore Default Workspaces to reset all non-user-defined workspaces.

To delete a workspace:

- Choose Delete Workspace from the Workspace menu to delete a saved workspace. You can't delete the workspace you're currently using. You must first pick a different one before choosing Delete Workspace.

Digital Imaging Essentials

To understand Photoshop's place in the pantheon of design software, and why it behaves the way it does, it's important to understand the framework it's working within.

This chapter focuses on key concepts for editing digital images: the implications of document size, the importance of resolution, and the distinction between the various Photoshop color modes. You'll also learn how the Info and History panels can help keep you informed and allow you the freedom to experiment and to recover from mistakes.

For basic digital image concepts, such as the difference between pixels and vectors, the math behind bit depths, and which file formats to use, see Appendix A, "Image Basics" online.

Adjusting Resolution and Image Size

Image resolution is measured in pixels per inch (ppi). For best results, images should contain the minimum resolution needed to obtain quality output from your target output device, at the desired output size. Because high-resolution photos contain more pixels, and therefore finer details, than low-resolution photos, they have a larger file size, take longer to render onscreen, require more processing time to edit, and are slower to print. There are no "right" resolutions, only *appropriate* resolutions—those that are neither too high nor too low.

The Image Size dialog (choose Image > Image Size) shows the relationship between image size and resolution (**FIGURE 3.1**). With the Resample option off, you can change either the dimensions *or*

the resolution—Photoshop adjusts the other value to preserve the total number of pixels. Increase the resolution and you reduce the width and height and vice versa: Increase the width and height and you reduce the resolution.

With Resample selected, you are changing an image's pixel dimensions. If you reduce (*downsample*) the number of pixels, information is deleted from the image. If you *upsample*, new pixels are added. Resampling, particularly upsampling, can result in reduced image quality.

Together, the width, height, and the resolution determine the file size of the image. The width and height also determine the base size at which an image is placed into another application like InDesign or Illustrator. For best print quality, change the dimensions or resolution first, without resampling. Then resample only if you have substantially more pixels than you need.

With Resample selected, you can change print dimensions and resolution independently (and consequently change the total number of pixels in the image) (**FIGURE 3.2**).

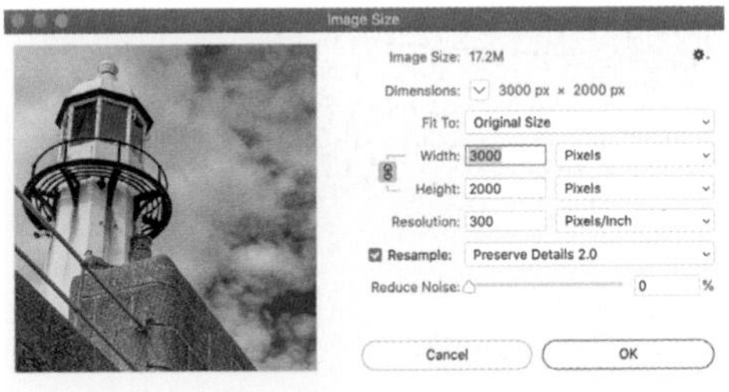

FIGURE 3.1 The Image Size dialog

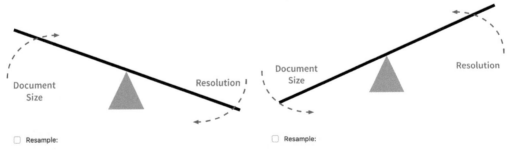

FIGURE 3.2 The resolution seesaw

Interpolation Methods

In a menu at the bottom of the Image Size dialog, Photoshop offers seven interpolation methods to determine how pixels are added or deleted when you resample an image (**FIGURE 3.3**).

Because each method has its strengths and weaknesses, Photoshop indicates its "best use." For example, Nearest Neighbor (Hard Edges) is good for resizing graphics and screenshots because they tend to have hard edges. It's not a good choice for photographs as they may have edges that aren't as well defined.

The Automatic option selects the best interpolation routine based on what Photoshop determines is the content of the image.

You can specify a default interpolation method in general preferences (Ctrl/Command+K) (**FIGURE 3.4**).

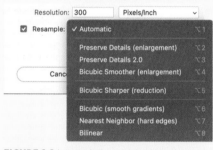

FIGURE 3.3 Interpolation methods in the Image Size dialog

FIGURE 3.4 Setting the default interpolation method

To change image size with Resampling on:

1. Choose Image > Image Size.

2. Select Resample and choose an interpolation method.

3. To constrain proportions, make sure the link icon (🔗) is unbroken. This keeps the width and height proportional.

4. Enter values for Width and/or Height. Optionally, choose Percent to enter values that are percentages of the current dimensions. The new file size is shown at the top of the dialog, with the old file size in parentheses.

Check Current Image Size

Keep track of your file sizes to make sure the files aren't becoming bigger than they need be. You can view the image file size in the Status bar at bottom left of the application window. The first number is the flattened size (more on this in Chapter 6); the second number is the uncompressed file size including any layers or channels. If you're opening a single-layer image, both numbers will start out the same. You can also click and hold to view an image's dimensions (**FIGURE 3.5**).

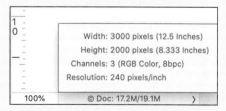

FIGURE 3.5 View the document size(s) and dimensions on the Status bar.

TIP If an image has layers with styles applied, click the gear icon to select Scale Styles if you want to also scale the effects in the resized image.

TIP For best results when downsampling to produce a smaller image, apply a Smart Sharpen filter afterwards.

Optimum Print Resolution

Print resolution is measured in ink dots per inch, also known as dpi. Printer resolution (dpi) is distinct from, but related to, image resolution (ppi). The optimum print resolution depends on how you are printing the document. For output to a desktop inkjet printer, an appropriate resolution is between 240 and 300 ppi. For commercial printing, the resolution is most likely to be 300 ppi, but you should ask your print shop what resolution they recommend for their press.

Screen Resolution

Monitor resolution is expressed in pixels. If your monitor resolution and your image's pixel dimensions are the same size, the image fills the screen at 100%. Just how big 100% looks depends on the image's pixel dimensions, the monitor size, and the monitor resolution setting.

For web or screen images, 72 ppi is often recommended. This is true up to a point, but what really matters is the pixel dimensions of the image. So what is the right number of pixels?

This depends on how the image will be used on screen (**FIGURE 3.6**). If you are preparing the image for a website, you need to estimate how large the average user's browser window will be, then calculate how much of that window you want the image to fill.

> **TIP** For statistics on browser display sizes visit www.w3schools.com/browsers/browsers_display.asp.

FIGURE 3.6 The same image at three sizes viewed in a large web browser (1920 x 1080)

Resolution and Image Dimensions

Resolution can be a hard concept to grasp with a fair bit of jargon to mystify new users. Let's look at some key terms:

- The **pixel count** (pixel dimensions) of an image is calculated by multiplying its height and width values (as in 3000 x 2000 pixels equals 6 megapixels).

- The **resolution** of an image (the pixel density) is measured in pixels per inch or ppi (as in 250 ppi).

- **Resampling** is the process of changing a file's pixel dimensions. Upsampling adds pixels; downsampling deletes pixels.

VIDEO 3.1

Understanding the Relationship Between Document Size and Image Resolution

Channels

Every Photoshop image has one or more *channels* that store information about the color in the image. The number of channels depends on the color mode. Channels in color images are actually grayscale images that represent each of the color components of an image. For example, an RGB image has separate channels for red, green, and blues color values.

The Channels panel lists all channels in the image (**FIGURE 3.7**). At the top of the list is the composite channel, which shows how the image looks in its normal state with all the channels visible. You can view any combination of channels for your image. Click the Visibility icon (the eyeball) next to the channel to show or hide that channel. (Click the composite channel to show all default color channels.)

Individual channels display in grayscale because they are easier to edit that way, but for RGB, CMYK, or Lab images, there is the option to view the individual channels in color. We recommend taking a quick look at this because it makes it easier to understand how channels work, then going back to viewing channels in grayscale.

FIGURE 3.7 The Channels panel and menu for an RGB document

To show color channels in color:

- Choose Edit/Photoshop > Preferences > Interface and select Show Channels In Color (**FIGURE 3.8**).

TIP To change the channel thumbnail size, choose Panel Options from the Channels panel menu.

To select a channel:

- Click its name. Shift-click to select (or deselect) multiple channels.

To edit a selected channel:

1. Choose a painting or editing tool.
2. Paint in white to add the selected channel's color.
3. Paint in black to remove the channel's color.
4. Paint in gray to add the channel's color at a lower intensity. You can paint on only one channel at a time.

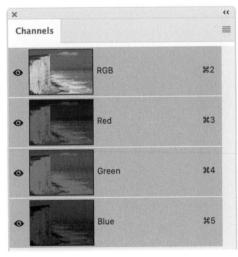

FIGURE 3.8 Viewing channels in color

Color Modes

An image's color mode determines how its colors combine based on the number of channels in a color model. In practical terms, Photoshop has five color modes.

RGB color mode

RGB is king of the color modes. It is the default mode for new images, and the mode in which digital cameras save your photos. It is the only mode in which all the tool options and filters are accessible; and the mode of choice for export to the web, mobile devices, video, and for printing on most inkjet printers. Not to mention, RGB has fewer channels than CMYK mode (so your computer uses less memory), while offering a wider range of colors (gamut).

RGB assigns a brightness value or level to each pixel. In 8-bit images, the levels range from 0 to 255 for each of the RGB (red, green, blue) components of a color image. When the values of all three colors are equal, the result is a neutral gray. When the values are all 255, the result is pure white; when they are 0, the result is pure black (**FIGURE 3.9**).

The exact range of RGB colors varies, depending on your monitor and the working space setting that you specify in the Color Settings dialog.

CMYK color mode

CMYK is the color mode used in color printing. Each pixel has a percentage value for each of the process inks: cyan, magenta, yellow, and black. The lightest (highlight) colors have small percentages of ink; the darker (shadow) colors have higher percentages. In CMYK images, pure white, or the absence of ink, is when all four components have values of 0% (**FIGURE 3.10**).

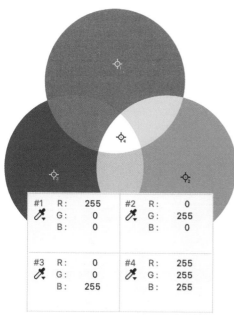

#1	R:	255	#2	R:	0
	G:	0		G:	255
	B:	0		B:	0
#3	R:	0	#4	R:	255
	G:	0		G:	255
	B:	255		B:	255

FIGURE 3.9 The RGB color model with color samplers showing levels

The exact range of CMYK colors varies, depending on the printing press and printing conditions, and on the CMYK working space that you specify in Color Settings.

Grayscale color mode

An 8-bit grayscale image has up to 256 shades of gray. Every pixel has a brightness level ranging from 0 (black) to 255 (white). Like CMYK values, grayscale values are measured as percentages, specifically of black ink (0% is equal to white, 100% to black).

Bitmap color mode

In Bitmap mode, the pixels are either 100% black or 100% white. No layers, filters, or adjustment commands are available. To convert a file to this mode, you must first convert it to Grayscale mode (**FIGURE 3.11**).

#1	C:	100%	#2	C:	0%
	M:	0%		M:	100%
	Y:	0%		Y:	0%
	K:	0%		K:	0%
#3	C:	0%			
	M:	0%			
	Y:	100%			
	K:	0%			

FIGURE 3.10 The CMYK color model with color samplers showing ink percentages

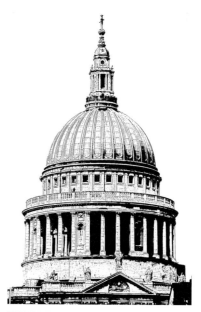

FIGURE 3.11 A bitmap image where all pixels are either black or white

Indexed color mode

Indexed color mode produces an image with up to 256 colors. A file in Indexed color mode contains just one channel and has no layers. When you convert an image to Indexed color, Photoshop creates a color lookup table (CLUT), which stores and indexes the image's colors. If a color in the image is not in the table, Photoshop chooses the closest color or simulates that color with available colors using dithering. Limited editing is available in indexed mode. If you need to edit an indexed color image, convert it temporarily to RGB mode (**FIGURE 3.12**).

TIP With RGB color mode, you can preview how an image will look when output from a particular device. To turn on this CMYK preview, choose View > Proof Setup Working CMYK, then View > Proof Colors (FIGURE 3.13).

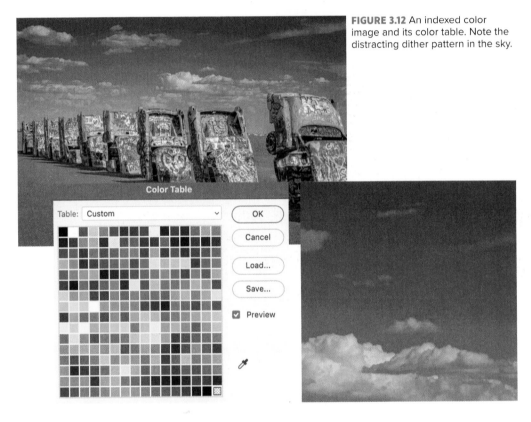

FIGURE 3.12 An indexed color image and its color table. Note the distracting dither pattern in the sky.

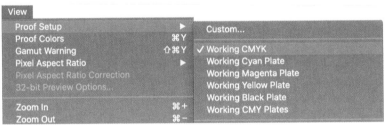

FIGURE 3.13 Specify how to preview your output with Proof Setup.

Conversion Between Color Modes

The world of graphic arts and design is a generally peaceful place populated by reasonable people who are supportive and happy to share information. Conflict is rare and typically everyone gets along. Except, that is, when it comes to converting between color modes, or more specifically when it comes to the most frequent mode conversion: RGB to CMYK.

Converting an RGB image into CMYK creates a color separation to make negatives and plates for commercial color printing. The four-color process requires four separations: cyan, magenta, yellow, and black (CMYK).

On one level, converting from RGB to CMYK mode is perfectly simple:

1. Choose Image > Mode > CMYK Color, or for more control, choose Edit > Convert To Profile.

2. A warning dialog will tell you that you are converting to a specific CMYK working profile: Confirm this is what you want by clicking OK, and you have a CMYK image (**FIGURE 3.14**).

But this is the tip of the proverbial iceberg. While some staunchly defend this workflow, others (us included) think that it is a bad idea (most of the time).

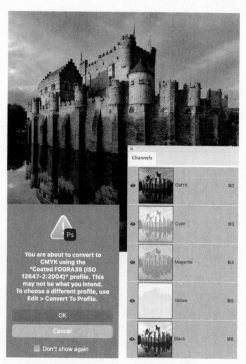

FIGURE 3.14 After converting to CMYK, the image has four channels, which can be made into separations at the time of printing.

Converting from RGB to CMYK is irreversible and results in lost information and, possibly, in noticeable color shifts. If your image starts out in RGB, it's best to do your edits in RGB and then convert to CMYK at, or near, the end of your workflow. If you're using a color managed workflow this conversion can take place outside of Photoshop, when an InDesign or Illustrator document containing placed Photoshop images is exported to PDF. You can also preview or soft proof an image to see how it will look when reproduced on a particular output device:

- Choose View > Proof Setup Working CMYK.

- Choose View > Proof Colors to turn on the CMYK preview (see Chapter 19 for more details).

We don't want to scare you. Some mode conversions are necessary and benign, like moving between Indexed color to RGB for editing flexibility, or moving from RGB or CMYK to Lab and back again. But in general, avoid multiple conversions between color modes. Never convert RGB images to CMYK if they are to be viewed onscreen. And if converting to CMYK in Photoshop, speak to your print service provider about which specific CMYK profile to use (they can provide you with one) for the specific printing conditions of the job.

The Info Panel

The Info panel displays a wealth of useful information, such as the color values of the cursor, and other details (**FIGURE 3.15**):

- Whether a color is out of the printable CMYK color gamut (an exclamation point appears next to CMYK values)

- The X and Y coordinates of your cursor

- The width (W) and height (H) of a marquee when using the Marquee tool

- The angle of rotation of the crop marquee when using the Crop tool

- The original X and Y coordinates, the change in X (DX), the change in Y (DY), the angle (A), and the length (D) of the line as you drag with the Line, Pen, or Gradient tools, or when moving a selection

- The percentage of the scale in width (W) and height (H), the angle of rotation (A), and the angle of horizontal skew (H) or vertical skew (V) when using the Free Transform tool

- The before-and-after color values for the pixels beneath the cursor and beneath any color samplers when making color adjustments (for example, when using Levels)

- Hints for using the selected tool when the Show Tool Hints option is enabled

- A variety of status information, depending on which option are selected

Both the First and Second Color Readouts let you choose a color mode—for example, to see hypothetically what the color values would be if the RGB image was converted to CMYK.

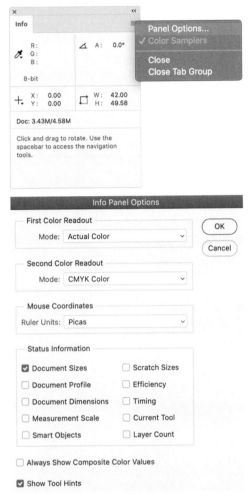

FIGURE 3.15 The many options available via the Info panel

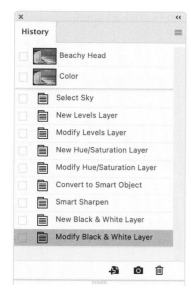

FIGURE 3.16 The History panel with the starting snapshot and a named snapshot at the top of the panel. Listed beneath are the history states.

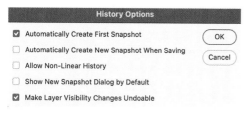

FIGURE 3.17 Choose whether to work using a nonlinear history.

Multiple Undo and the History Panel

If you plan on never making any mistakes in Photoshop, you can skip this section. But for mere mortals, it's essential.

The History panel keeps a running list of the *states*, or edits, made to the open document, from the Open (unedited) state of the document at the top of the list to the most recent state at the bottom. In the current work session, you can selectively restore your document to one of its previous states (**FIGURE 3.16**).

To restore a document to an earlier state:

- Click an earlier state to restore the document to that point. Recent states may become dimmed and unavailable when you do this, depending on whether the panel is in Linear or Nonlinear mode (**FIGURE 3.17**).

In *Linear mode,* you can restore your document to an earlier state with a clean break. If you start editing from an earlier state, all subsequent (dimmed) states are lost.

Nonlinear mode is the more flexible, but can be confusing. In *Nonlinear mode*, if you select or delete an earlier state, subsequent states are retained. Your next edit becomes the latest state on the panel, incorporating the earlier stage of the image plus your newest edit. The states in between are preserved, and if you change your mind, you can click an intermediate state to resume editing from that point.

Each open document has its own list of states. Because it can take a chunk of memory to record multiple history states, you might want to manage the number of history states you save.

To manage history states:

1. Go to Edit/Photoshop > Preferences > Performance.

2. Under History & Cache, choose a History States value (the default is 50). If you run out of states, the oldest are deleted to make room for new ones.

TIP **To step back one history state, press Ctrl/Command+ Z. To step forward one state, press Ctrl+Shift+Z/Command+Shift+Z. To toggle the last state (to compare a before and after) press Ctrl+Alt+Z/Command+Option+Z.**

You can't change the order of history states, but you can delete them.

To delete a history state:

Do one of the following:

- Right-click a state, choose Delete from the context menu, then click Yes in the alert dialog. That state and all subsequent states are deleted.

- Drag the state to the Delete Current State button to bypass the alert.

- Click a state, then Alt/Option-click the Delete Current State button as many times as needed to delete previous states in reverse order without an alert.

When you purge or clear states from the History panel, the most recent state is left as the only state remaining. All snapshots are preserved.

To purge or clear the History panel:

Do one of the following:

- Choose Edit > Purge > Histories to clear all states from your History panel for all currently open documents to free up memory.

- Right-click any state and choose Clear History to clear all states from the History panel for just the current document. This command doesn't free up memory, but it can be undone.

Crop and Straighten Images

A good crop is often one of the most important—and simplest—things you can do to improve an image. The edges of an image make up its frame, and your choice of framing strengthens what you think is important. If you're the one capturing the image, you can crop in camera, but even with the best forethought in the world, you may change your mind when you see the image on screen. You may need to crop to remove distracting elements, to empha-size what's important, or to repurpose an image for a specific page or screen size and shape.

The Crop tool provides real-time feedback to help you visualize the cropped result. You can make permanent changes that delete the cropped pixels or nondestruc-tive changes that retain the cropped pixels in case you need to revert to them. You can also straighten crooked images and option-ally plug any gaps with Content-Aware Fill, which samples pixels from adjacent areas. The related Perspective Crop tool can be used to fix perspective distortions.

In This Chapter

Cropping a Photo

Cropping is one of the most basic types of image manipulation and potentially one of the most impactful. It can remove unwanted objects or distractions from the edge of the frame, change the image's aspect ratio, and improve its overall composition.

1. Select the Crop tool (C) (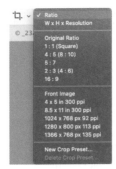) from the toolbar. Crop handles display on the edges of the photo.

2. Draw a crop area or drag the corner and edge handles to define the crop boundaries of the photo (**FIGURE 4.1**).

3. (Optional) On the Options bar, choose a ratio or physical size for the crop. You can enter values in the Width and Height fields, choose a preset, or define your own preset values for later use (**FIGURE 4.2**).

4. (Optional) Choose to display overlay guides, such as Rule Of Thirds, Grid, and Golden Ratio, while cropping. To cycle through all the options, press O (**FIGURES 4.3** and **4.4**).

5. Click the gear menu on the Options bar to specify additional crop options as desired (see the sidebar "Additional Crop Options").

6. Press Enter/Return or click the checkmark on the Options bar to crop the photo (**FIGURE 4.5**).

TIP Make sure to provide enough "breathing space" around your subject when composing photos. Its always easier to take extra stuff away than it is to add something you don't have.

TIP If you have an active selection, you can also crop an image using the Crop command. Use a Selection tool to select the part of the image you want to keep, then choose Image > Crop.

FIGURE 4.1 A crop in progress with the Rule of Thirds overlay shown

FIGURE 4.2 Crop options: Ratio and size presets

FIGURE 4.3 Overlay options

FIGURE 4.4 Use the crop overlays to help compose the final image. While you can't go wrong with Rule of Thirds, different types of images lend themselves to different overlays. Press O to cycle through overlays.

FIGURE 4.5 Use the Options bar to Reset, Cancel, or Commit to the crop.

Cropping with Content-Aware Fill

Cropping implies making the image smaller, but you can also "crop" to extend an image. Using Content-Aware technology, Photoshop will intelligently fill the new areas. This works on a single-layer image only, however.

1. On the Options bar, select Content-Aware. The default Crop rectangle expands to include the whole image.

2. Straighten or rotate the image or expand the canvas beyond its original size (**FIGURE 4.6**).

3. Press Enter/Return, or click on the checkmark in the Options bar to commit to the crop.

FIGURE 4.6 Using Content-Aware Fill to add more sky to the top of the image

Additional Crop Options

Beyond overlays and crop dimensions, you can customize your crop using a few choices from the Option bar's gear menu (**FIGURE 4.7**).

FIGURE 4.7 Additional crop options

Use Classic Mode makes the Crop tool behave the way it did in previous versions of Photoshop.

Auto Center Preview keeps the preview in the center of the canvas.

Show Cropped Area displays the area that is cropped. If you disable this option, only the final area is previewed.

Enable Crop Shield overlays the cropped areas with a tint. You can specify a color and opacity.

Delete cropped pixels means cropped pixels are permanently removed, recoverable only if its Undo state is still available. Disable this option for a nondestructive crop that hides but does not delete the pixels outside the crop boundaries. Switch to the Move tool and drag to move the image around within the cropped area. Or to view the hidden pixels, choose Image > Reveal All.

 VIDEO 4.1
Cropping and Straightening an Image

Aspect Ratio

Aspect ratio is the relationship between the width and the height of an image. It is expressed as two numbers separated by a colon, such as 3:2. The first number refers to the horizontal side of the image, and the second number refers to the vertical side. For example, 3:2 indicates a horizontal image, whereas 2:3 refers to a vertical image. Aspect ratio is sometimes also expressed as a decimal number, such as 1.50 (the long side divided by the short side). Aspect ratio does not represent the physical size of an image, nor its dimensions in pixels. For example, a 3:2 aspect ratio could translate to an image that is 3 inches wide and 2 inches high, or 18 inches wide and 12 inches high.

The most common aspect ratios are:

1:1 (1.00): Instagram made 1:1 popular by initially requiring every photo to be square. It now accommodates different aspect ratios, but square images dominate.

5:4 (1.25): This aspect ratio is common when printing 8x10- and 16x20-inch images.

3:2 (1.50): Most DSLR, mirrorless, and point-and-shoot cameras have 3:2 sensors, regardless of the sensor size. The 3:2 aspect ratio was popularized by 35mm film and is the most common one in photography.

4:3 (1.33): Medium format, Micro Four Thirds, most smartphones, and some point-and-shoot cameras have 4:3 sensors.

16:9 (1.78): The most common video format. Using this aspect ratio can give your images a cinematic feel.

3:1 (3.0): This is the aspect ratio commonly used for panoramas.

The right aspect ratio is ultimately a matter of personal taste (**FIGURE 4.8**). You will encounter some images that just don't easily conform to a standard aspect ratio and that you will want to crop in a freeform way. Sticking to standard aspect ratios, however, where possible will make it easier to find standard size frames for your images and also to incorporate your images into page layouts.

FIGURE 4.8 Four common aspect ratios indicated with color guides and the resulting cropped images all shown at the same scale

Straightening Images

We're often surprised by how photos we thought were straight when we took them look slightly crooked on screen. Whether this is down to user error or the optics of our lenses, thankfully it's easy to fix and can really improve an image. And when you straighten a photo while cropping, Photoshop resizes the canvas automatically to accommodate the rotated pixels.

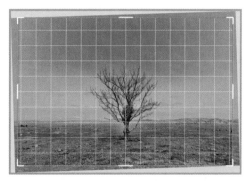

FIGURE 4.9 Cropping and straightening using the grid.

To straighten an image while cropping:

1. Select the Crop tool (C). Crop handles display on the edges of the photo.

2. Start dragging from one of the corner handles. Hover your cursor outside the crop boundary until it changes to a curved double-ended arrow.

3. Move the cursor just outside the image and drag to rotate the image. A grid displays inside the crop box, which you can use for reference while you rotate the image behind it (**FIGURE 4.9**).

4. Press Enter/Return or click the checkmark on the Options bar to confirm the crop.

FIGURE 4.10 Draw a line along the horizon with the Straighten tool.

To straighten an image with the Straighten tool:

1. Select the Crop tool.

2. Click Straighten on the Options bar.

3. Draw a reference line with the Straighten tool along the horizon or an edge to straighten the image along it (**FIGURE 4.10**).

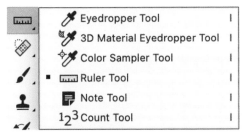

FIGURE 4.11 The Ruler tool

FIGURE 4.12 For a quick-and-dirty batch scan and crop, use File > Automate > Crop and Straighten Photos.

To straighten an image with the Ruler tool:

1. Select the Ruler tool (**FIGURE 4.11**). (If necessary, click and hold the Eyedropper tool to reveal the Ruler.)

2. In the image, drag across an element that should be horizontal or vertical.

3. On the Options bar, click Straighten Layer. Photoshop straightens the image.

4. Use the Crop tool or Content-Aware Fill to eliminate any transparent areas along the edges.

Crop and straighten scanned photos

You can place several photos on a scan bed and scan them in one pass to create a single image file. The Crop and Straighten Photos feature will then create separate files from the multiple-image scan.

TIP For best results, have a solid white background behind the photos to make it easier for Photoshop to isolate each photo. Work with images that have clearly defined edges, and leave about ⅛ inch (3 mm) between the images.

To crop and straighten scanned photos:

1. Open the scanned file of the images you want to separate.

2. Choose File > Automate > Crop And Straighten Photos. Photoshop does its magic and each image opens in its own window (**FIGURE 4.12**).

Perspective Cropping

Use the Perspective Crop tool to fix images with keystoning (distortion that can happen when a building is shot from below) or other distortions that can happen with a wide-angle lens. This tool is also great for straightening the edges of photographed artwork. (A related tool, Perspective Warp, is discussed in Chapter 15.)

1. Select the Perspective Crop tool.

2. Draw a marquee around the distorted image.

3. When you release the marquee, Photoshop converts it to a bounding box with handles. Drag the handles to match the edges to the object you want to correct (**FIGURE 4.13**).

4. Press Enter/Return.

Cropping with the Trim Command

By now you may be getting the impression that there are multiple ways to do things in Photoshop, and cropping is no exception. The Trim command provides another option. It trims surrounding transparent pixels or background pixels of a color you specify.

Choose Image > Trim, and in the Trim dialog, select an option (**FIGURE 4.14**):

- **Transparent Pixels** to trim away transparency at the edges of the image

- **Top Left Pixel Color** to remove an area defined by the color of the top-left pixel

- **Bottom Right Pixel Color** to remove an area defined by the color of the lower-right pixel

- **Top**, **Bottom**, **Left**, or **Right** to remove one or more areas of the image to trim away.

FIGURE 4.13 Cropping with the Perspective Crop tool

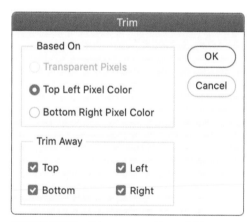

FIGURE 4.14 The Trim command is useful for trimming away a border of transparent pixels.

TIP If trimming away the transparent pixels still leaves a transparent space around the image, it's because the pixels aren't fully transparent; this is common with drop shadows.

Rotating or Flipping an Entire Image

The Image Rotation command rotates or flips an entire image (**FIGURE 4.15**). The equivalent commands on the Edit > Transform submenu transform only the selected layer (see Chapter 15).

To rotate an image, choose Image > Image Rotation and one of the following commands from the submenu:

- **180°** rotates the image by a half-turn.

- **90° Clockwise** rotates the image clockwise by a quarter-turn.

- **90° Counter Clockwise** rotates the image counterclockwise by a quarter-turn.

- **Arbitrary** rotates the image by the angle you specify.

- **Flip Canvas Horizontal** or **Flip Canvas Vertical** flips the image along the corresponding axis.

TIP While we're on the topic of rotation, let's clarify the difference between rotating the image and rotating the *view* of an image. Image Rotation modifies the file information, whereas Rotate View nondestructively rotates the image for viewing.

FIGURE 4.15 The result of flipping the canvas horizontally

Changing the Canvas Size

The canvas is the editable area of an image. You can make it bigger or smaller using the Canvas Size command. Canvas size is distinct from image size, and it can be bigger or smaller than the image itself. Increasing the canvas adds space around an image. If the image has a transparent background, the added canvas will be transparent. If it doesn't, the added canvas is a solid color. Decreasing an image's canvas size crops into the image—yet another way to crop! If your layer is a Normal (as opposed to Background) layer, you can use the Move tool to adjust the crop of the image within the canvas.

To change the canvas size:

1. Choose Image > Canvas Size.
2. In the Canvas Size dialog that opens, do one of the following (**FIGURE 4.16**):
   - ▸ Enter the canvas dimensions in the Width and Height fields. As well as entering the dimensions in inches, centimeters, or pixels, you can also express this as a percentage of the current canvas size.
   - ▸ Select Relative, and enter the amount you want to add or subtract from the current canvas size.
3. Click a square to anchor the image to the new canvas. The size change will happens on the side(s) opposite the selected square or from the center if the center square is selected.
4. Choose an option from the Canvas Extension Color menu. If the layer is unlocked, the canvas extension will be transparent.

You can also use the Crop tool to resize the image canvas, too.

To resize the canvas using the Crop tool:

1. Drag the crop handles outwards to enlarge the canvas. Hold Alt/Option to enlarge from all sides.
2. Press Enter/Return to confirm.

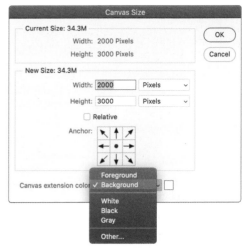

FIGURE 4.16 The Canvas Size dialog and Canvas Extension color options. Note that the Canvas Extension Color options are applicable for only images with a Background layer.

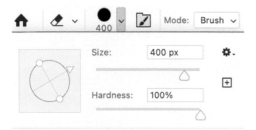

FIGURE 4.17 Choose the history state that you want to erase to and then paint on the image.

FIGURE 4.18 The Eraser settings on the Options bar

FIGURE 4.19 Erasing to a history state. Interesting as this may be, there are easier ways—see Chapter 7.

Erasing Parts of an Image

As well as cropping out unwanted parts of an image at the edges of a frame, you can also delete parts of the image within the area that you retain. Photoshop has several erasing tools that share the same tool space: the Eraser, the Background Eraser, and the Magic Eraser.

The **Eraser** erases to the background color if you're on a background or locked layer; otherwise, the pixels are erased to transparency. There's also an option to erase to a state on the History panel.

To erase to a history state:

1. Make a change to the image. In this example, we chose Image > Adjustments > Black And White.

2. Undo the change by pressing Ctrl/ Command+Z.

3. Select the Eraser tool (E) (🖊), and on the Options bar, check Erase To History.

4. On the History panel, click in the left column of the dimmed history state (Black & White) (**FIGURE 4.17**) so that the History Eraser icon appears in that column.

5. Paint on the image to restore areas to the specified history state. Adjust the size and settings of the Eraser as necessary (**FIGURES 4.18** and **4.19**).

Erasing with the Background Eraser

The **Background Eraser** erases similar pixels to transparency as you drag, overriding any lock transparency settings. It samples the color in the center, or hotspot, of the brush, and deletes that color wherever it appears inside the brush.

1. Select the Background Eraser (🧽) (E).

   ▸ Optionally, add a Color Fill layer beneath the image layer.

   ▸ Adjust size and settings of the brush.

   ▸ Choose a sampling option. If the background is the same color, choose Sampling Once. (**FIGURE 4.20**)

   ▸ Choose the Sampling Limits: Discontiguous lets you paint between strands of hair or branches (**FIGURE 4.21**).

   ▸ Choose a tolerance setting: a higher number removes more colors.

2. Paint around the edge to remove the background color, revealing transparency or the color fill layer beneath. Vary the brush size as go. (**FIGURE 4.22**)

3. Switch to regular Eraser to clean up any areas away from the edges. (**FIGURE 4.23**)

Erasing with the Magic Eraser

The **Magic Eraser** changes similar pixels to transparent. If the transparency is locked, the pixels change to the background color. If you're working on a Background layer, it is converted to a normal layer and all similar pixels change to transparent.

The Background Eraser and Magic Eraser are older tools and better, faster, and nondestructive results can be achieved using the Select and Mask workspace. (See Chapters 6–8 and 16 on layers, layer masks, and retouching.)

FIGURE 4.20 Choosing a sampling option

FIGURE 4.21 Setting the sampling limits

FIGURE 4.22 Paint around the edges of the subject. Vary the brush size (press [for a smaller brush, or] for a larger brush), sampling limits, and tolerance as necessary.

FIGURE 4.23 Switch to the regular Eraser to paint out areas away from the edges.

5

Selections

Because so much of what we do in Photoshop starts with a selection, using the Selection tools is a fundamental Photoshop skill. A selection isolates one or more parts of your image for manipulation while leaving the unselected areas untouched. Making a selection is a means to an end and not the end itself. Once you've made a selection, it exists as potential for how you can alter the image.

There are several ways to make selections, using tools and menus. Every selection method has its pros and cons. None will work in every instance. The key is knowing when to choose one tool, or combination of tools, over another.

In This Chapter

Marquee Tools

These tools let you select rectangular and elliptical shapes. You can also select 1-pixel-wide rows and columns.

To make a rectangular or elliptical selection:

1. Click the layer you want to target.

2. Click and hold the current Marquee tool (M) (**FIGURE 5.1**), then choose Rectangular Marquee Tool to make rectangular selections, or Elliptical Marquee Tool to make elliptical selections.

3. (Optional) On the Options bar, adjust the Feather and Anti-alias settings (see Chapter 3).

4. For the Style setting, choose Normal (**FIGURE 5.2**). You will control the marquee size and proportions by dragging.

5. Drag diagonally over the area you want to select. For a square or circle, hold Shift while you drag (**FIGURE 5.3**).

TIP Hold Alt/Option to draw a rectangular or elliptical selection from the center.

TIP For more precision than dragging, set Style to Fixed Ratio and specify a height-to-width Fixed Ratio or Fixed Size and specify the marquee's height and width. Then, click and drag over the canvas area you want to select.

TIP When using Fixed Ratio or Fixed Size, click the arrows icon (⇄) to swap the width and height values.

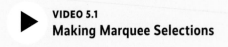

▶ **VIDEO 5.1**
Making Marquee Selections

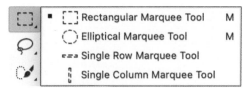

FIGURE 5.1 The Marquee tools

FIGURE 5.2 Options for the Elliptical Marquee tool

FIGURE 5.3 Use the Elliptical Marquee tool (holding Alt/Option to draw from the center) to make selections of circular or oval subjects.

Select All

The most "macro" selection you can make is to select all pixels of a layer within the canvas boundaries.

To select all:

1. Target the layer in the Layers panel.

2. Choose Select > All (Ctrl/Command+A). A marching ants selection border will now appear around the bounds of your image.

FIGURE 5.4 In this example, a single column of pixels is selected (from the center of the image), and then transformed to create an abstract image.

To select a single row or column of pixels:

Practical uses of the single row or single column marquee tools are few and far between, and you may have a long and illustrious Photoshop career without ever using them, but here's a fun case study that can result in a pleasing abstract image.

1. Click the layer you want to target.

2. Choose either the Single Row or Single Column Marquee tool.

3. Click anywhere on the canvas to make the selection.

4. Copy the selection to a new layer (Ctrl/Command+J).

5. Choose Edit > Free Transform, and stretch the pixels across the whole canvas (**FIGURE 5.4**).

Anti-aliasing

Because pixels are square, anti-aliasing is needed to smooth selection edges that would otherwise be jagged. It does so by softening the color transition between edge pixels and background pixels.

Anti-aliasing is available for the Lasso, Polygonal Lasso, Magnetic Lasso, Elliptical Marquee, and Magic Wand tools. It's one of those options that you want turned on 98.5% of the time.

You must specify the Anti-alias option before using the tools. After a selection is made, you can't add anti-aliasing.

Lasso Tools

The Lasso comes in three flavors: "regular," Polygonal, and Magnetic (**FIGURE 5.5**).

FIGURE 5.5
The Lasso tools

Use the Lasso tool to make loose selections or to refine selections made with other tools, such as the Magic Wand or Quick Selection tool. Use the Polygonal Lasso tool to select a straight-sided subject. Working with a pressure sensitive tablet, it's possible to make accurate freeform selections with the Lasso tool; working with a mouse or trackpad, it's best to restrict the tool to making loose selections or to clean up rough selections (**FIGURE 5.6**). As for the Magnetic Lasso, which snaps to a selection edge, we never use it—everything it does can be done more easily with other selection tools.

To make freeform selections:

1. Select the layer on which you want to make the selection.
2. Select the Lasso tool (L or Shift+L).
3. (Optional) Set values for Feather and Anti-alias on the Options bar.
4. Drag around an area on the layer. When you release the drag, the open ends of the selection join automatically.
5. (Optional) To add to the selection, Shift-drag around the area to be added.
6. (Optional) To subtract from the selection, Alt/Option-drag around the area to be removed.

VIDEO 5.2
Making Lasso Selections

FIGURE 5.6 Use the Lasso tool to make rough selections, such as this free-form lasso selection around the dark clouds in preparation for removing them with Content-Aware Fill (see Chapter 13).

To make straight-sided selections:

1. Select the Polygonal Lasso tool (L or Shift+L).
2. Click to create straight-sided segments.
3. To close the selection, position the cursor over the starting point and click when a closed circle appears beside the cursor (**FIGURE 5.7**). Alternatively, you can simply release the mouse.

FIGURE 5.7 The Polygonal Lasso tool is a good choice for selecting straight-sided subjects.

TIP Toggle between the Lasso tool and the Polygonal Lasso by holding Alt/Option.

TIP To constrain the selection to multiples of 45°, hold Shift while clicking.

TIP To delete the last segment while using the Polygonal Lasso tool, press Backspace/Delete.

Automated Selection Tools

The automated selection tools—Magic Wand, Quick Selection, and Object Selection—share the same tool space and single-key shortcut (W). Before making a manual selection, it's worth trying these tools first. Even if they don't make a perfect selection, they usually give you a starting point.

Magic Wand

The Magic Wand has been a part of Photoshop from the very beginning, and since that time it has been responsible for the selection of more than a gazillion pixels. With the Magic Wand tool, you can click on an area of the image to select all adjacent pixels of the same (or a similar) shade or color.

It's intuitive, and it really does feel like magic—when it works. The thing is, it won't always work. Don't let your Magic Wand become a Tragic Wand. If it ain't working, try another tool. There may be easier, faster, ways that result in better quality selections.

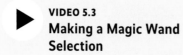

VIDEO 5.3
Making a Magic Wand Selection

To make a selection with the Magic Wand tool:

1. Select the Magic Wand tool (W or Shift+W).

2. Set the Sample Size (the number of pixels the Magic Wand should evaluate) in the Options bar. To avoid sampling an errant, non-representative pixel color, rather than use Point Sample, choose 3 By 3 Average or 5 By 5 Average (**FIGURE 5.8**).

FIGURE 5.8 The Sample Size options

3. On the Options bar, set the Tolerance, which determines how similar in tone or color the other pixels need to be in order to be selected. A higher number results in a bigger selection; the default value is 32 (**FIGURE 5.9**).

FIGURE 5.9 The results of clicking on the poppy with the Magic Wand with Tolerance set to 32 (left) and 64 (right).

4. Still on the Options bar, turn on the Anti-alias option, which creates a smoother-edged selection—and is nearly always a good idea.

5. (Optional) Turn on the Contiguous option on the Options bar. With Contiguous checked, the Magic Wand selects only adjacent pixels of the same brightness. When not checked, all pixels in the image with the same brightness values are selected (**FIGURE 5.10**).

FIGURE 5.10 Using the Magic Wand with Contiguous not selected, results in all similar colors in the image being selected.

6. Click on a representative area of the image to make a selection with the Magic Wand.

TIP To select similar colors on all visible layers, turn on the Sample All Layers option on the Options bar. To select colors on just the current layer, make sure this option is off.

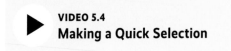

VIDEO 5.4
Making a Quick Selection

Quick Selection tool

The Quick Selection tool lets you "paint" a selection using an adjustable elliptical brush. As you drag, the selection expands outward, automatically finding the defined edges in the image.

To make a selection with the Quick Selection tool:

1. Select the Quick Selection tool (W or Shift+W).

2. On the Options bar, choose New to make an initial selection, Add To to add to an existing selection, or Subtract From to remove areas from an existing selection.

3. (Optional) To change the brush options for the Quick Selection tool, click the Brush Options on the Options bar (**FIGURE 5.11**).

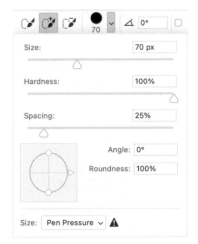

FIGURE 5.11 The Brush options for the Quick Selection tool

4. Turn on the Enhance Edge option to create a smoother selection edge (**FIGURE 5.12**). We've never found a reason not to use this.

FIGURE 5.12 Quick Selection tools

5. Drag over the part of the image you want to select (**FIGURE 5.13**).

FIGURE 5.13 The Quick Selection tool makes easy work of this kind of selection because the edges are well defined against the background.

TIP To subtract from a selection, toggle to Subtract mode by holding Alt/Option, then drag over the existing selection.

TIP Both the Add and Subtract modes help "teach" the Quick Selection tool what to pay attention to when finding edges. In some cases edge detection can improve by not just dragging within the subject in Add mode, but also by dragging outside the subject in Subtract mode.

TIP To quickly change the brush size or hardness, press the Right Bracket (]) key to increase the brush size or the Left Bracket ([) key to decrease the brush size. Press Shift+] to increase the hardness or Shift+[to decrease hardness.

TIP To factor in the pixels on all layers, rather than just the currently selected layer, turn on Sample All Layers before painting your selection.

Object Selection tool

The Object Selection tool, the newest of the automated selection tools, makes selecting one or more objects in an image easier and is an efficient way to create a starting point for a more accurate selection.

To make a selection with the Object Selection tool using the Object Finder:

1. Turn on the Object Finder option on the Options bar; the Object Selection tool automatically finds objects in your image (**FIGURE 5.14**).

FIGURE 5.14 Turn on Object Finder to automatically identify objects in the image.

2. Highlight an object by hovering your cursor over it. A color overlay (blue by default) will indicate the found objects (**FIGURE 5.15**).

FIGURE 5.15 Using artificial intelligence, Object Finder detects the objects in the image. When the cursor hovers over the sunflower, it shows a colored overlay.

3. Click the object to instantly select it.

TIP To add to the selection, hold Shift and click on another object.

TIP To subtract a currently selected object, hold Alt/Option and click on the object.

As great as the Object Finder is, there will be times when it doesn't know exactly what you want. For example, using the same image, to select just the center of the sunflowers, we need to be more specific. You can make the selections yourself in either Rectangle or Lasso mode. Drag to draw a rectangle or freeform area to select the object inside the defined region. Once the selection has been made, as well as marching ants, a color overlay appears when you move inside the selected area(s).

To make a selection with the Object Selection tool without Object Finder:

1. Select the Object Selection tool.
2. Choose the mode (Rectangular or Marquee).
3. For a sharper edge, choose the Hard Edge option.
4. To include pixels from all layers in the selection, choose Sample All Layers.
5. Draw a marquee over the object you want to select (**FIGURE 5.16**).

If you need to refine your first selection, you can switch modes (between Rectangular and Lasso) and use the Object Selection tool in Add, Subtract, or Intersect mode. See "Working with Selections."

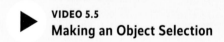
▶ **VIDEO 5.5**
Making an Object Selection

FIGURE 5.16 Using the Object Selection tool in Rectangular mode, a selection is dragged over the center of the flower. The Object Selection tool finds the object within the selection area.

The Refresh Icon

When you select the Object Selection tool, the Refresh icon (♻) next to the Object Finder option spins. This indicates that Photoshop is analyzing the image looking for objects; wait until it stops spinning for the best results.

The Object Finder will automatically refresh and reanalyze the image anytime you make a change. But you can also refresh it manually by clicking the Refresh icon.

Deselecting and Reselecting Selections

What goes up must come down and what gets selected must also be deselected. If you plan on using the selection again, you can save it, so that you don't need to repeat steps—but we'll get to that in Chapter 8.

To deselect selections:

Do one of the following:

- Choose Select > Deselect (Ctrl / Command+D).

- Right-click and choose Deselect.

- Or, if you're using the Rectangle Marquee tool, the Elliptical Marquee tool, or the Lasso tool, click anywhere in the image *outside* the selected area.

TIP Perhaps the number one reason for Photoshop not behaving the way you expect is that you have a forgotten or hidden selection—even a tiny one. This can flummox even advanced users. As a first troubleshooting step, choose Select > Deselect (Ctrl/ Command+D) to see if that fixes the problem.

Offering more than just a simple undo, Photoshop will remember your last selection, and you can recall this selection even after you've done other things to the image.

To reselect the most recent selection:

Do one of the following:

- Choose Select > Reselect (Ctrl+Shift+D/ Command+Shift+D).

- Click the selection step in the History panel.

Select Subject

Available on the Options bar and via the Select menu when you're clicked on an image layer and have an automated selection tool chosen, Select Subject uses Adobe's artificial intelligence to analyze the image and decide what to select.

If the image has a clearly discernible subject, it can work well. But using the Object Selection tool increases your chances of getting a good result because it gives Photoshop an idea of what you're after. (**FIGURE 5.17**)

FIGURE 5.17 The difference between Select Subject (top) and the Object Selection tool—drawing a marquee around the right swan (bottom)

Working with Selections

Having looked at some basic ways of making selections, let's turn our attention to some ways in which you can modify those selections. You can transform selections, move them, hide their edges, add to or subtract from them in various ways, copy them to another document, and more.

If you need to make minor tweaks to the shape of a selection, then Transform Selection is a useful tool. Note that this changes the selection *outline* only, not the contents of the selection.

To transform a selection:

1. With a selection active, choose Select > Transform Selection.

2. Pull on any other transform handles to reshape the selection (**FIGURE 5.18**).

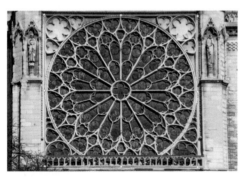

FIGURE 5.18 Distorting an elliptical selection using Transform Selection.

TIP To distort a selection, hold Ctrl/Command while dragging a transform handle.

TIP To transform a selection from the center, hold Alt/Option while dragging a transform handle.

TIP To change the contents of a selection, choose a command from the Edit > Transform submenu instead of choosing Select > Transform Selection.

To move a selection edge:

1. Select any selection tool (except Quick Selection or Object Selection).

2. Position the cursor inside the selection and drag.

TIP To move a selection while drawing it, keep the mouse button down, then hold the spacebar.

TIP To constrain the move direction to multiples of 45°, begin dragging, and then hold Shift as you continue to drag.

TIP To move the selection in 1-pixel increments, press an arrow key.

TIP To move the selection in 10-pixel increments, hold Shift, and use an arrow key.

TIP To move the pixels of the selection, rather than the selection edge, use the Move tool.

Sometimes selection edges can be visually distracting, and you'll want to hide them.

To hide selection edges (aka marching ants):

Do one of the following:

- Choose View > Extras to show or hide selection edges, grids, guides, target paths, slices, annotations, layer borders, count, and Smart Guides all at once (**FIGURE 5.19**). (You can specify what constitutes an "extra" by choosing View > Show > Show Extras Options.)

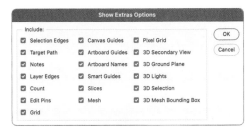

FIGURE 5.19 The Show Extras menu and Show Extras Options

- Choose View > Show > Selection Edges to toggle the view of the selection edges of the current selection only. The selection edges reappear when you make a different selection.

You can refine your initial rough selections by adding to them or subtracting areas from them. With the Quick Selection tool, doing so is as easy as choosing Add mode (its default behavior) or Subtract mode, and then painting over the areas you want to add to or remove from the initial selection. For other selection tools it takes an additional step.

To add to a selection:

1. While using any selection tool (except Quick Selection), select Add To Selection on the Options bar or hold Shift. A plus sign appears beside the cursor when you're adding to a selection.

2. Drag the tool to select the area you want to add to the selection (**FIGURE 5.20**).

FIGURE 5.20 Adding to a selection

To subtract from a selection:

1. Select Subtract From Selection on the Options bar or hold Alt/Option. A minus sign appears beside the cursor when you're subtracting from a selection.

2. Drag the tool to select the area you want to remove from the selection (**FIGURE 5.21**).

FIGURE 5.21 Draw a rectangular marquee over the raspberry while holding down Alt/Option to subtract it from the original selection.

Intersect is more useful when you're making selections based on luminance or color values (see Chapter 8) than selecting objects. But here's a simple example.

To intersect with a selection:

- Select Intersect With Selection on the Options bar or hold Shift+Alt/Option. An X appears beside the cursor when you're selecting an intersected area (**FIGURE 5.22**).

FIGURE 5.22 Start out with all berries selected. Then, to intersect the selection, leaving just the bottom halves of each selected, draw a rectangular marquee while holding Alt/Option+Shift.

Before manually adding to or subtracting from a selection in any way, set the Feather and Anti-aliasing values on the Options bar to the same settings used in the original selection (**FIGURE 5.23**).

FIGURE 5.23
The Selection options from left to right: New, Add to Selection, Subtract from Selection, Intersect with Selection

Sometimes selecting the opposite parts of an image from those you want is easier than selecting the parts you do want. In those cases, you can use the Inverse option. Suppose you had an object on a solid-colored background. You could easily select the background with the Magic Wand tool, and then invert the selection to become the object (**FIGURE 5.24**).

To inverse a selection:

1. Make a selection using any selection method.
2. Choose Select > Inverse (Ctrl+Shift+I/ Command+Shift+I).

To move selected pixels:

1. Switch to the Move tool (V).
2. Drag from within an active selection to a new position. If you selected multiple areas, all areas will move as you drag.

To copy selected pixels:

1. Switch to the Move tool (V).
2. Hold Alt/Option as you drag from within an active selection to a new position.

FIGURE 5.24 It's often easier to start with the opposite of what you want. In this example, because the sky is almost flat, it is easy to select with the Magic Wand or by choosing Select > Sky. Then, choose Inverse to select the buildings.

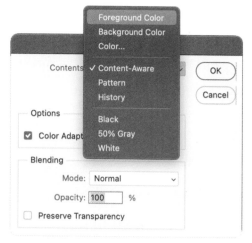

FIGURE 5.25 Deleting a selection: before and after

FIGURE 5.26 Choose a fill option.

To copy a selection to another document:

1. Switch to the Move tool (V).

2. Drag from within the selection to the other document's tab or document window, pause, and then drop it.

To delete a selection:

- Make sure the selection is active and that you have the right layer selected.

- Press Backspace/Delete to delete the selection and replace it with transparency (**FIGURE 5.25**)—just so long as the layer is not a Background layer.

To fill a selection:

1. Make sure the selection is active and that you have the right layer selected.

2. Choose Edit > Fill (Shift+F5) to open the Fill dialog.

3. Choose the type of fill you want from the Contents menu (**FIGURE 5.26**).

4. Click OK.

TIP To quickly fill a selection with the Foreground color, press Alt+Backspace/Option+Delete.

TIP To quickly fill a selection with the Background color, press Ctrl+Backspace/Command+Delete.

Modifying Selections

With the evolution of the Select and Mask dialog (see Chapter 8) and its more inter-active interface, the options on the Modify menu—Border, Smooth, Expand, Contract, and Feather—are not as essential as they once were. Nevertheless, the old-school methods are still useful.

Border lets you select a width of pixels inside and outside an existing selection edge.

To modify the border of a selection:

1. Make a selection with one of the selection tools.
2. Choose Select > Modify > Border.
3. Enter a value between 1 and 200 pixels in the Width field, and click OK (**FIGURE 5.27**).

Smooth can reduce the bumpiness of a rough selection.

To smooth a rough selection:

1. Make a selection with one of the selections tools.
2. Choose Select > Modify > Smooth.
3. Enter a value for the Sample Radius (**FIGURE 5.28**).

To expand or contract a selection:

1. Make a selection with one of the selections tools.
2. Choose Select > Modify > Expand or Select > Modify > Contract.
3. In the Expand By or Contract By field, enter a specified number of pixels by which to expand (push out from its current edge) or contract (pull in from its current edge) the selection (**FIGURE 5.29**).

FIGURE 5.27 The result of creating a border selection (in this case, of 30 pixels)

FIGURE 5.28 Close-up of the result of smoothing a rough selection by 3 pixels

FIGURE 5.29 From left to right: close-up of a selected red object, the selection expanded, the same selection contracted, a selection border.

FIGURE 5.30 The result of copying a feathered selection to a new layer and then hiding the original layer

To feather a selection:

1. Make a selection with one of the selections tools.

2. Choose Select > Modify > Feather.

3. In the Feather Radius field, enter a value by which to soften the selection edges. The effect of the feathering becomes visible only after you move, cut, copy, or fill the selection (**FIGURES 5.30**).

TIP While you can add feathering when using the Marquee tools and the Lasso tools, it's preferable to start with a neutral (unfeathered) selection and add feathering as needed. Once a selection edge has been feathered, it can't be unfeathered.

TIP Small selections with a large feather radius may be so faint that their edges are invisible and you see this warning message: "No pixels are more than 50% selected." You can either decrease the feather radius or increase the size of the selection. Or live with the selection knowing that you won't be able to accurately see the selection edges.

Growing Selections

Both Grow and Similar select a range of pixels up to the current Tolerance setting of the Magic Wand tool.

To grow a selection:

1. Make an active selection.

2. Choose Select > Grow to include all adjacent (contiguous) pixels.

3. Alternatively, choose Select > Similar to extend a selection to include similar (contiguous and noncontiguous) pixels (**FIGURE 5.31**).

TIP To increase a selection in increments, choose either the Grow or Similar command more than once.

FIGURE 5.31 The original selection is on the left. Choosing Grow extends the selection to adjacent pixels; choosing Similar includes noncontiguous pixels.

6

Layers

Almost everything we do in Photoshop uses layers, and layers are so fundamental that it's strange to remember they didn't exist in early versions of Photoshop. When layers were introduced in version 3, it was like when silent movies became talkies. A missing dimension was added, and everyone wondered how on earth we ever managed without them.

Layers let you manipulate your images in ways big and small, simple and fantastic. Some things wouldn't be possible without layers, but even those that would, would be a lot harder. Layers allow you to keep the various elements of a composition separate, so that you can edit them independently. You can use layers to move elements, to add text or graphic shapes, or to combine images with opacity and blending effects.

Think of layers as a stack. You build your composition by stacking layers from the bottom up. And you can change the order of the layers by moving them up or down.

In This Chapter

About Layers

By default, a new image has a single layer, listed as Background on the Layers panel. By adding layers of various kinds above the Background layer, you can build complexity and flexibility into your documents.

For example, you can add text on one layer and perhaps a shape or some imagery surrounded by transparency on another. You can add, delete, hide, or change the order of layers when needed. Layers can be edited independently, giving you complete control of your composition. It is this independence of layers that makes them such a key part of a nondestructive workflow (see the sidebar "Techniques for Nondestructive Editing").

Unlike the Background layer, which is always opaque, locked, and at the bottom of the stack, a layer can contain partially or fully transparent areas and can be moved upward or downward in the stack. You can see through any transparent parts of a layer to the layers below. Photoshop uses a checkerboard pattern to represent transparent areas.

The Layers panel (Window > Layers) lists all layers, layer groups, and layer effects in an image (**FIGURE 6.1**). Here, you can show and hide layers, create new layers, and work with groups of layers. You can access additional commands and options by clicking the Layers panel menu at the upper-right of the panel.

TIP You can change the size of layer thumbnails by choosing Panel Options from the Layers panel menu.

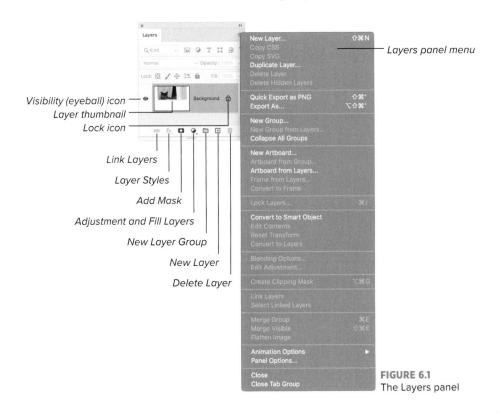

Layers panel menu

Visibility (eyeball) icon
Layer thumbnail
Lock icon
Link Layers
Layer Styles
Add Mask
Adjustment and Fill Layers
New Layer Group
New Layer
Delete Layer

FIGURE 6.1
The Layers panel

Types of Layers

While it's not common to use all types of layers in a single composition, that's what we're doing in **FIGURE 6.2** so you can understand what's happening on the Layers panel.

A: An **image layer** of texture at reduced opacity.

B: A **layer group** comprised of three type layers.

C: The different layer groups are **color coded** for easy identification.

D: The down pointing arrow indicates a **layer clipped** to the layer below (lips).

E: A Hue/Saturation **adjustment layer** changes the color of the lips.

F: The badge indicates that this layer has been converted to a **Smart Object**.

G: A **vector shape layer**.

H: A solid color **fill layer**.

I: A **warped type layer**.

J: The **FX badge** indicates a layer effect (here a drop shadow) applied to the layer.

K: A **layer mask** applied to the Zip layer hides selected parts of that layer.

L: The layers with a **link icon** have been linked, so that they can be moved as a single item.

M: A **Smart Filter** is applied to give a painterly look to the lips.

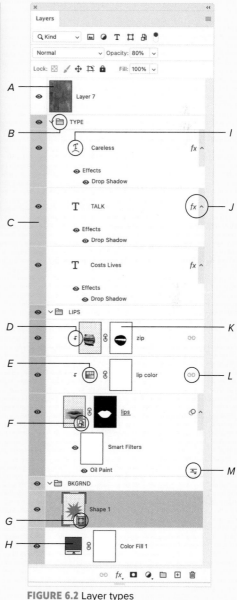

FIGURE 6.2 Layer types

The Background Layer vs. Regular Layers

When you create a new image with a white or colored background, the layer is called Background. An image can have only one Background layer. You can't move it or adjust its blending mode, and it is always opaque. However, you can easily convert a Background to a normal layer. Less frequently, you may want to convert a layer to a Background layer. You can do that, too.

When you create a new document with transparent Background Contents, the image does not have a Background layer.

To convert a Background to a layer:

Do one of the following:

- Click the lock icon (🔒) to the right of the Background thumbnail.

- Double-click the Background to open the New Layer dialog and change any layer options as desired.

To convert a layer to a Background:

- Choose Layer > New > Background From Layer. The layer goes to the bottom of the layer stack and any transparent pixels are converted to the Background color.

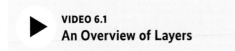

VIDEO 6.1
An Overview of Layers

Creating Layers

When you create a new layer, it appears either above the selected layer or within the selected layer group in the Layers panel. The new layer starts out as transparent, as indicated by the checkerboard button shown on its thumbnail in the Layers panel. You can add as many layers as you like, limited only by your storage space and available system memory.

To create a new layer:

- Click the New Layer button (⊞) at the bottom of the Layers panel. The new layer appears above this selected one. The new layer will have the default blending mode of Normal and the default setting of 100% for both Opacity and Fill.

To choose options for a layer as you create it:

1. Alt/Option-click the New Layer button, or press Ctrl+Shift+N/ Command+Shift+N. The New Layer dialog opens.

2. In the New Layer dialog, choose your desired options and name the layer (**FIGURE 6.3**).

3. (Optional) Choose a nonprinting color to identify the layer.

FIGURE 6.3 Naming a layer and choosing options

To copy a selection to a new layer:

1. Create a selection.

2. Choose Layer > New > Layer Via Copy (Ctrl/Command+J).

To cut a selection and convert it to a new layer:

1. Create a selection.

2. Right-click in the document and choose Layer Via Cut (Ctrl+Shift+J/ Command+Shift+J). The selected area of the original layer will become transparent (**FIGURE 6.4**); if you cut pixels from the Background, the selected area is filled with the Background color.

To duplicate a layer or layer group:

Do one of the following:

- On the Layers panel, click a layer or a layer group, or Ctrl/Command-click multiple layers, then press Ctrl/Command+J.

- On the Layers panel, drag a layer, a layer group, or the Background layer over the New Layer button. The duplicate will appear above the one you dragged.

TIP When you duplicate a layer, any masks and effects on that layer are also duplicated.

FIGURE 6.4
The difference between Layer via Copy (middle) and Layer via Cut (bottom)

Copying and Pasting Layers

You can copy and paste layers within a document and between documents:

- **Paste** (Edit > Paste or Ctrl/Command+V), which creates a duplicate layer in the center of the destination document. The new layer includes any layer masks, vector masks, and layer effects.

There are some variants of the Paste command on the Paste Special menu:

- **Paste In Place** pastes the copied layer(s) into the destination document in the same relative position as the original document.

- **Paste Into** or **Paste Outside** pastes a copied selection into or outside another selection in any image. The source selection is pasted onto a new layer, and the destination selection border is converted into a layer mask.

To paste one selection into another:

1. Make a selection in the source document and choose Edit > Copy (Ctrl/Command+C) (**FIGURE 6.5a**).

2. Move to the destination document and select the area you want to paste into. (**FIGURE 6.5b**)

3. Choose Edit > Paste Special > Paste Into. (**FIGURE 6.5c**)

4. (Optional) Position the pasted layer within the layer mask that is created.

a

b

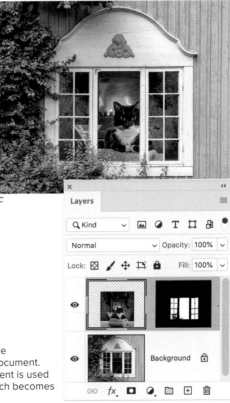

c

FIGURE 6.5 Pasting a selection from one document into a selection in another document. The selection in the destination document is used as a mask for the pasted selection, which becomes its own layer.

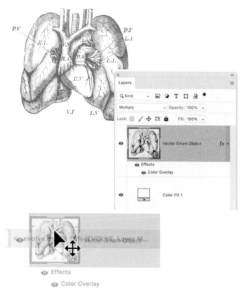

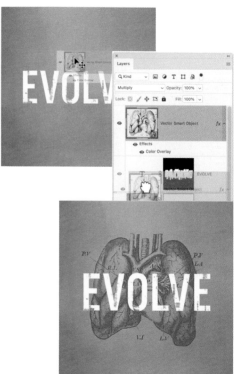

You can also copy layers or layer groups to another image by dragging.

To copy a layer or layer group between images by dragging:

1. Open the source and destination images.

2. From the Layers panel of the source image, select one or more layers or a layer group.

3. Drag the layer or layer group from the Layers panel, hold it over the tab of the destination image until its window comes forward, and then finish by dropping it inside the destination window (**FIGURE 6.6**). In the Layers panel, the duplicate layer or group will land above the active layer in the destination image.

4. (Optional) Hold Shift as you drag to make the layer or group land in the same location in the destination image that it occupied in the source image (assuming both images have the same pixel dimensions.)

TIP To exclude layer properties, such as blending modes, when copying a single layer, choose Select > All, then Edit > Copy. Move to the destination document and choose Edit > Paste in the destination image.

TIP If you drag a file from your desktop onto an open image, Photoshop creates a Smart Object layer. You can change this behavior in General Preferences by deselecting Always Create Smart Objects When Placing.

TIP Depending on your color management settings, you may see a warning message about color profiles when copying and pasting between documents. The safest response is to choose Preserve Embedded Profiles (see Chapter 11).

FIGURE 6.6 Dragging a layer from one document to another. The layer lands above the currently active layer, so you may need to change the stacking order, by dragging the layer up or down.

Layers: A Spotter's Guide

Here's a brief description of the different layer types and how to recognize them.

The **Background layer** (🔒) is how most Photoshop images begin their life. The Background layer differs from other layers in that you can't change its stacking order, blending mode, or opacity. However, you can convert a Background to a regular layer by double-clicking its thumbnail and naming it something else.

Layer groups (📁) are folders into which you can put related layers, enabling you to conceptually order your layers as well as keep things neat and tidy. Click the chevron to the left of the group folder to expand or contract the view of the group contents.

Layer effects (*fx*) allow you to quickly and nondestructively add shadows, glows, bevels, and the like. Layer effects are linked to the layer contents.

Type layers (**T**) are created automatically when you use the Type tool. The type remains editable, and you can add layer effects to your type as well as apply transformations (except Perspective and Distort). If you want to apply filters to type, first convert the type layer to a Smart Object.

Shape layers (▱) can be created by using the Pen tool or the Shape tools, which create vector shapes—useful when you need simple graphic shapes with crisp edges.

Adjustment layers (◑) are used to apply color and tonal adjustments in a nondestructive way.

Layer masks (◖) are like alpha channels but attached to specific layers. They allow you to determine what parts of a layer are shown and what parts are masked or hidden.

Clipping masks (ꜜ) allow you to restrict the influence of specific layers. Typically, a layer affects everything that is below it in the layer stack. Using a clipping mask, layers can be clipped by a base layer so that they affect only that base layer (**FIGURE 6.7**).

Smart Object layers (🗗) are embedded files that maintain a link to the original data, which means you can nondestructively transform their content.

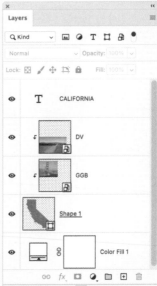

FIGURE 6.7 The two image layers are clipped to the shape beneath. An alternative way to achieve the same result would be to put both layers in a group and add a vector mask to the group.

Selecting Layers

When a layer or layer group is selected, it has a highlight, which by default is gray. A selected layer is called the *active* layer, and its name appears in the title bar of the document window.

For tasks like painting or making color and tonal adjustments, you can work on only one layer at a time. For tasks such as moving, aligning, transforming, or applying styles from the Styles panel, you can select and affect multiple layers at a time:

- To select multiple contiguous layers, click the first layer, and then Shift-click the last layer.

- To select multiple noncontiguous layers, Ctrl/Command-click them in the Layers panel.

- To select all layers, choose Select > All Layers.

- To deselect all layers, click in the Layers panel below the background or bottom layer, or choose Select > Deselect Layers.

If you're not getting the results you expect, you may not have the correct layer selected. Check the Layers panel to make sure that you're on the right layer for what you're trying to do.

It can sometimes be faster to select a layer or layers directly from the document window.

To select a layer from the document window:

- Right-click the image and choose a layer from the context menu. The context menu lists all the layers that contain pixels beneath the cursor (**FIGURE 6.8**).

FIGURE 6.8 Selecting a layer by right-clicking

You can also use the Auto-Select option.

To auto-select a layer:

1. Select the Move tool (V).

2. On the Options bar, select Auto-Select, then choose Layer from the menu (**FIGURE 6.9**).

FIGURE 6.9 Auto-selecting a layer

3. Click in the document on the layer you want to select. The topmost layer containing pixels under the cursor is selected.

To auto-select a layer group:

1. Select the Move tool (V).

2. On the Options bar, select Auto-Select, then choose Group from the menu.

3. Click in the document on the content you want to select. The topmost group containing pixels under the cursor is selected. If you click an ungrouped layer, it becomes selected.

TIP As an alternative to choosing Auto-Select on the Options bar, toggle to Auto-Select by holding Ctrl/Command. It's a more fluid way of working.

Moving, Aligning, and Transforming Layers

Showing the boundary or edges of the content in a layer can help you move and align the content. You can also display the transform handles for selected layers and groups, making it easier to resize or rotate them.

To show a layer's edges:

1. Select the layer.

2. Choose View > Show > Layer Edges to show the edges of the content in a selected layer (**FIGURE 6.10**).

To move the content of layers:

1. Select the layer(s).

2. With the Move tool (V), reposition the content by dragging.

TIP You can also use the arrow keys to nudge the content by 1 pixel or by 10 pixels (if you hold Shift).

It's a fundamental design principle that the elements of a composition should some-how be aligned—whether it's with each other or with the canvas. The alignment options in Photoshop let you align the content of two or more selected layers by their edges or centers. You can also distribute the spacing among three or more selected layers.

To change layer order in the layer stack:

- To change the layer or group stacking order, drag them up or down in the Layers panel, or choose Layer > Arrange and choose a command from the submenu.

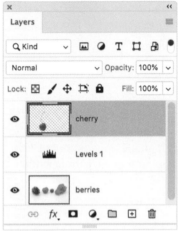

FIGURE 6.10 Showing the edges of the selected layer

FIGURE 6.11 Rotating a layer. Note the angle of rotation in the cursor.

a

b

c

FIGURE 6.12 The results of aligning three shape layers by the top edges (a), bottom edges (b), and vertical centers (c)

To rotate a layer:

1. To rotate a layer choose Edit > Transform > Rotate Or Edit > Free Transform. A bounding box appears.

2. Move the cursor outside of the bounding box (it becomes a curved, two-sided arrow ↰), and then drag (**FIGURE 6.11**). Hold Shift while dragging to constrain the rotation to 15° increments.

3. To commit the rotation, press Enter/Return, or click the checkmark on the Options bar. To cancel, press Esc, or click the Cancel icon on the Options bar.

To align layers:

1. Select the Move tool (V).

2. Select two or more layers.

3. Click one of the Align buttons on the Options bar, or choose Layer > Align (**FIGURE 6.12**).

TIP To align layers to a selection, create a selection before clicking the Align buttons.

To distribute space among layers:

1. Select the Move tool (V).

2. Select three or more layers.

3. Click one of the Distribute options on the Options bar.

To see the transform handles of a selected layer:

1. Select the Move tool (V).

2. On the Options bar, select Show Transform Controls.

Export layers

You can export and save layers as individual files using a variety of formats, including PSD, JPEG, and TIFF. Layers are named automatically as they are saved. This can be a time-saving feature when working on a series of images that all have the same treatment.

To export layers:

1. Choose File > Export > Layers To Files.

2. (Optional) By default, Photoshop saves the generated files in the same folder as the source file, but you can choose a destination folder.

3. Set the other options such as name prefix, file format, and whether you want to only export visible layers (**FIGURE 6.13**).

Changing layer opacity

The Layers Opacity option controls whether a layer is semitransparent or opaque (**FIGURE 6.14**).

You can also change the opacity of the selected layer or layer group via the keyboard. Choose the Move tool or a Selection tool, then press a number between 0 and 9 (e.g., 2 for 20%) or type a percentage (such as 57 for 57%). Press 0 to set the opacity to 100% or 00 to set it to 0%.

Related to opacity are the layer blending modes, which are the subject of Chapter 10. The default blending mode and opacity is Normal at 100%, which means that the colors of the layer will not interact with the colors of the layers below.

FIGURE 6.13 Exporting layers as individual PSD (native Photoshop) files

FIGURE 6.14 The Opacity and Fill options on the Layers panel. Opacity affects the transparency of the whole layer, including any layer effects. Fill does not affect the opacity of any layer effects. (See Chapter 10.)

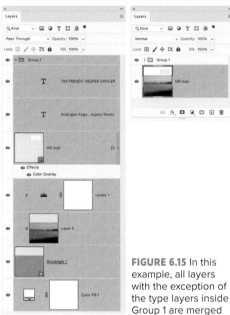

FIGURE 6.15 In this example, all layers with the exception of the type layers inside Group 1 are merged into a single layer.

Merging and Flattening Layers

Sometimes it's good to reduce the number of layers to keep down file size and minimize clutter on your Layers panel. There are several ways to do this. First and foremost, be cautious and don't commit to anything that will limit how you can edit the document in the future.

Merge Down

Merge Down merges the selected layer with the layer below. When you save a merged document, the layers are permanently merged; you can't revert back to the unmerged state.

To merge select layers:

1. Select the layers and/or groups you want to merge.

2. Choose Layer > Merge Layers (**FIGURE 6.15**).

To merge all visible layers and groups:

1. Hide any layers you don't want merged.

2. Choose Merge Visible from the Layers panel or the Layers panel menu.

> **TIP** You can merge an adjustment layer into an image layer, but you can't merge adjustment layers into one another. If you merge a shape layer, an editable type layer, or a Smart Object, it is rasterized into the underlying image layer.

Stamp multiple layers or linked layers

Rather than the nuclear option of merging all your layers, which runs counter to a nondestructive approach, you can keep your layers intact and merge your visible layers into a new layer above them. When you stamp multiple selected layers or linked layers this way, you create a new layer containing the merged content, while preserving the original, separate layers beneath. This is an effective way of consolidating your progress to a specific point while at the same time retaining the original layers in case you need to return to them.

To stamp visible layers:

1. Select the layers and/or groups you want to merge.

2. Hold Alt/Option and choose Layer > Merge Layers or use the shortcut Ctrl+Alt+Shift+E/ Command+Option+Shift+E (**FIGURE 6.16**).

Rasterize layers

If you want to paint directly on—or apply filters to—layers that contain vector data (such as type, shape layers, vector masks, or Smart Objects), you must first rasterize these layers to convert their contents to pixels.

To rasterize layers:

Do one of the following:

- Select the layers you want to rasterize, then choose Layer > Rasterize (**FIGURE 6.17**).

- Right-click to the right of the layer name and choose Rasterize from the context menu.

A word of caution: Once you rasterize a layer, you lose the benefits of scalability and editable type, so don't do this unless you absolutely must.

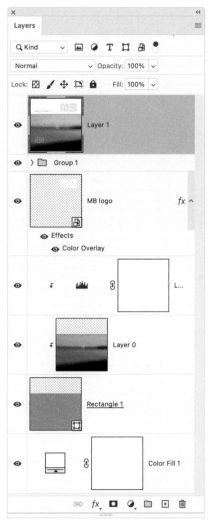

FIGURE 6.16 Choosing merge visible while holding Alt/Option retains the original layers below the merged result.

FIGURE 6.17 Rasterizing a Smart Object

FIGURE 6.18 Click Save a Copy to have access to all file types.

FIGURE 6.19 Choose JPEG from the Format menu.

Transparency Preferences

Typically, Photoshop represents transparency as a checkerboard to distinguish it from white. Most of the time this is useful, but occasionally the checkerboard becomes distracting. You can turn it off or change its size and color. Choose Edit/ Photoshop > Preferences > Transparency & Gamut (**FIGURE 6.20**).

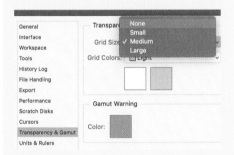

FIGURE 6.20 Choose the size of the transparency grid (checkerboard) in Transparency & Gamut Preferences.

Flatten layers

Flattening reduces file size by merging all visible layers into a single background and discarding hidden layers. Any transparent areas become white. This is a one-way ticket: When you save a flattened image, you can't revert back to the unflattened state. For that reason, only if you're 110% confident that your image is totally complete, that it will never under any circumstances require further editing, should you choose Layer > Flatten Image.

Yes, there are times when an image needs to be in a "flattened" state, without any layers—like putting an image on a webpage or social media post or sending a proof to a client. On such occasions, you save a flattened copy.

It's true that keeping your layers intact will mean a larger file size, but you should balance this with maximizing your editing flexibility. (Follow the housekeeping advice below to keep your files as organized and as lean as possible.) Keeping your layers means you'll always be able to make changes.

So, when you think of flattening always append the word *copy* to the filename.

To save a flattened copy:

1. Choose File > Save As (Ctrl+Shift+S/ Command+Shift+S). The Save As dialog opens.

2. Click Save A Copy (**FIGURE 6.18**).

3. Choose a file format from the Format menu. If you choose JPEG, a format that does not support layers, the Layers checkbox becomes deselected (**FIGURE 6.19**).

4. Click Save. The layered version remains open; the flattened version is saved to disk.

Layer Housekeeping

A tidy Layers panel is a sign of a tidy mind some wise person once said. Working fast and getting a little sloppy like a master chef tossing ingredients about can sometimes yield great results. But after you've paused for breath, go back, and clean up that mess. The next day, when you try to figure it out and possibly try to replicate what you did, you'll be glad you took the time to tidy. Imagine how glad you'll be when you open a file you created last week, last month, last year—or even last decade—to find it logically organized, and to know that you don't need a forensics team to figure out what's going on. Arrange your stuff logically. Leave a traceable path back to what did (**FIGURE 6.21**).

Organizing layers

There are a couple of simple things you can do for a more efficient workflow. Descriptive names make layers and groups easier to identify in the Layers panel. Keep in mind that you may revisit projects months or years from now. Or you may need to share a project with colleagues.

To add a descriptive layer name:

- Double-click the layer or group name in the Layers panel and type a new name.

It also helps to color code layers and groups, making it easier to find related layers in the Layers panel.

VIDEO 6.2
Managing Layers

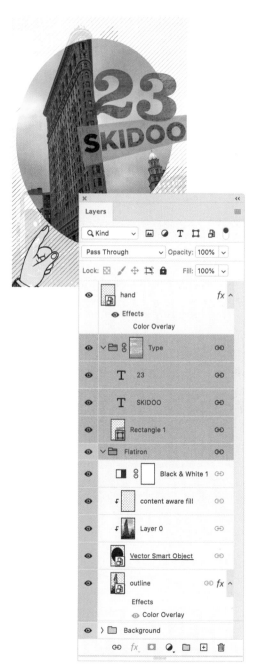

FIGURE 6.21 Organizing layers takes only a moment, and it will save time in the long run. Layers and groups are named, color coded, and, where appropriate, linked.

To color code layers:

- Right-click the layer or group and select a color.

Layer thumbnail display

You can control the visual feedback that you get from your layer thumbnails by setting a couple options.

To customize layer thumbnails:

1. Right-click any layer thumbnail and choose a size for all layer thumbnails.

2. Right-click again and choose the content shown in the thumbnail:

   - **Clip Thumbnails to Layer Bounds** means that only the layer content is shown.

   - **Clip Thumbnails to Document Bounds** means the layer content is shown relative to the canvas size.

These options as well as some others are also available in the Layers Panel Options dialog (**FIGURE 6.22**) from the Layers panel menu.

Lock layers

Layers can be locked fully or partially to protect their contents using the Lock Layers dialog. When partially locking a layer, you can choose **Lock Transparent Pixels** (to prevent editing outside of the opaque portions of a layer), **Lock Image Pixels** (to prevent painting on a layer), or **Lock Position** (to prevent a layer from being moved). When locked, a lock icon appears to the right of the layer name. The icon is solid when the layer is fully locked and hollow when partially locked.

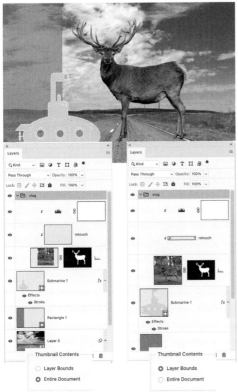

FIGURE 6.22 Layers panel options. The thumbnail contents are viewed relative to the entire document or to the layer bounds.

To fully lock layers:

1. Select a layer or multiple layers.

2. Choose Lock Layers from the panel menu.

3. Choose All (**FIGURE 6.23**). This will lock all linked layers if a single layer was selected.

To partially lock a layer:

1. Select a layer or multiple layers.

2. Choose Lock Layers from the panel menu.

3. Click one or more lock options in the Layers panel.

TIP For type and shape layers, Lock Transparency and Lock Image are selected by default and can't be deselected.

Filtering layers

Layer filtering is another useful housekeeping feature, especially if you're working on a complex document that contains many layers. The filtering options let you view a subset of layers based on name, kind, effect, mode, attribute, or color label.

To filter layers in the Layers panel:

1. Choose a filter type, such as Kind, from the menu at the top of the panel (**FIGURE 6.24**).

2. Select or enter the filter criteria.

3. Toggle the switch to turn layer filtering on or off.

To deactivate layer filtering, in the upper-right corner of the Layers panel, click the Layer Filtering On/Of button. (If you click the button again, the last settings are restored.)

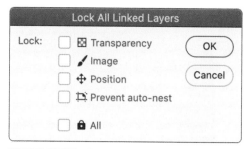

FIGURE 6.23 The locking options for layers

FIGURE 6.24 Choose the filtering method from the menu (Kind), a layer type icon (from left: Image Layers, Adjustment Layers, Type Layers, Vector Layers, Smart Objects), and click the toggle switch (far right) to turn filtering on and off.

TIP Click the Visibility (eyeball) column to show or hide specific layers or groups, or styles, so that you can more easily work on portions of your image independently.

TIP Alt/Option-click a Visibility icon to display just that layer or group. Alt/Option-click the same eyeball to restore the original visibility settings.

TIP Select just the layers that you want to view, then right-click the canvas and choose Isolate Layers to hide all other layers.

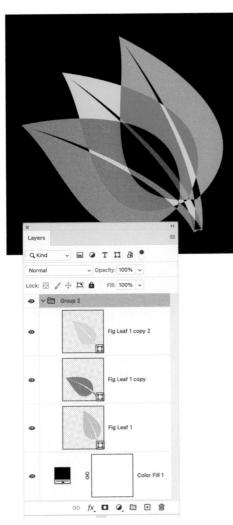

FIGURE 6.25 The three leaf shapes are combined in a layer group. Not only does this keep the Layers panel tidy, but in this case also allows for a transparency blending option to be applied to the leaves independently of the black background.

Layer groups

Layer groups cut down on clutter and arrange your layers in a logical order. You can move, transform, duplicate, restack, hide, show, lock, or change the opacity or blending mode of layer groups. You can put groups within groups. You can also apply attributes and masks groups to affect multiple layers simultaneously.

To create a layer group of existing layers:

1. Select multiple layers in the Layers panel.

2. Choose Layer > Group Layers (Ctrl/ Command+G), or drag the layers to the folder icon at the bottom of the Layers panel (**FIGURE 6.25**).

TIP In the Layers panel, click the arrow to the left of a group to expand and collapse that group.

TIP Hold Alt/Option and click the arrow next to a layer group to expand the layers along with any layer effects and Smart Filters.

TIP To drag multiple contiguous layers to another document, it's simpler to group them and then drag the group, rather than drag a multiple selection.

To create a new, empty layer group:

1. Click the layer above which you want the group to appear.

2. Click the New Group button. (To name the group as you create it, Alt/Option-click the New Group button or choose New Group from the panel menu.)

3. Drag the layer(s) to the group folder.

To move a layer into a group:

- Drag the layer to the group folder. If the group is closed, the layer goes to the bottom of the group. If the group is open, you can drop the layer anywhere in the group.

To ungroup layers:

1. Select the group.

2. Choose Layer > Ungroup Layers (Ctrl+Shift+G/Command+Shift+G).

Link and unlink layers

Linking layers is another way to establish a relationship between them. You can move or apply transformations to linked layers. Linked layers stay linked until you unlink them.

Linked layers and layer groups are very similar. With linked layers you can simultaneously reposition layers that aren't adjacent in a layer stack. For everything else, including moving and transforming, layer groups are preferable.

To link layers:

1. Select two or more layers or groups.

2. Click the Link icon [ᴇᴏ] at the bottom of the Layers panel.

To select all linked layers:

1. Select one of the layers.

2. Choose Layer > Select Linked Layers.

To unlink layers:

1. Select one or more linked layers.

2. Click the Link icon at the bottom of the Layers panel.

TIP To temporarily disable a linked layer, Shift-click the Link icon for the linked layer. A red X appears. Shift-click the Link icon to enable the link again.

Delete a layer or group

Removing layers you no longer need or that are abandoned experiments you'd like to forget can help tidy your Layers panel.

To delete layers:

1. Select the layers to remove in the Layers panel.

2. Do one of the following:

   ▸ To delete with a confirmation message, click the trashcan at the bottom of the Layers panel. (When you get tired of seeing this message, click Don't Show Again, or hold Alt/Option when you click the trashcan.)

   ▸ To delete hidden layers, choose Layers > Delete > Hidden Layers.

   ▸ To delete empty layers, choose File > Scripts > Delete All Empty Layers.

To delete all linked layers:

1. Select a linked layer.

2. Choose Layer > Select Linked Layers.

3. Delete the layers.

Techniques for Nondestructive Editing

Layers are an integral part of a nondestructive workflow. Nondestructive editing allows you to change an image without the changes being permanent or degrading the image quality. This maximizes your creative options, allowing you to experiment without fear of painting yourself into a corner. It's also more efficient because you can reuse the same elements in multiple documents.

You can work nondestructively in several ways:

Adjustment layers apply color and tonal adjustments without permanently changing pixel values. The edits are stored as instructions in the adjustment layer rather than applied directly to the image pixels. (See Chapter 9.)

Smart Objects contain one or more layers of content. What makes them smart is that you can transform (scale, skew, or reshape) the content without directly editing image pixels. Or you can edit the Smart Object as a separate image and when you have made your edits the changes update in place. Smart Objects also let you apply Smart Filters, so that you can later adjust or remove the filter effect. (See Chapter 14.)

Filters applied to Smart Objects become Smart Filters. You'll always be able to see and modify what filter was applied and what settings were used. Filter masks let you selectively conceal the effects of Smart Filters on Smart Object layers. (See Chapter 16.)

☑ Sample All Layers **Retouching on a separate layer.** Most retouching tools (Clone Stamp, Healing Brush, Spot Healing Brush, Content-Aware Patch, and Content-Aware Move) let you add your retouching to a separate layer. Be sure to select Sample All Layers on the Options bar. (See Chapter 13.)

Editing In Adobe Camera Raw preserves the original image data. When working with raw files, Adobe Camera Raw stores adjustments in separate sidecar files. (See Chapter 20, online.)

☑ Open in Photoshop as Smart Object **Open As Smart Object.** Choosing this option in the Adobe Camera Raw plug-in (ACR) lets you to edit Camera Raw settings at any time, even after you edit the file. (See Chapter 20, online.)

☐ Delete Cropped Pixels **Cropping nondestructively.** Deselecting Delete Cropped Pixels lets you restore the cropped area at anytime by choosing Image > Reveal All or by dragging the Crop tool beyond the edge of the image. (See Chapter 4.)

Layer masks and vector masks can be continuously edited without harming the pixels they hide. (See Chapter 7.)

Layer Comps

Designers often create multiple versions of a composition (or comps) to show clients. A layer comp is a snapshot of a state of the Layers panel. Using them you can create, manage, and view multiple versions of a layout in a single Photoshop file (**FIGURE 6.26**).

You can specify several options for layer comps upon creation:

- **Visibility:** Whether a layer is showing or hidden.
- **Position:** The layer's position in the document.
- **Appearance:** Whether a layer style is applied to the layer and the layer's blending mode.
- **Layer Comp Selection for Smart Objects:** When you select the Smart Object in the containing document, the Properties panel gives you access to the layer comps that were defined in the source file. This means you can change the state of the Smart Object at a layer level without editing the Smart Object.

To create a layer comp:

1. Choose Window > Layer Comps to open the Layer Comps panel.

2. Click the Create New Layer Comp button (+ icon) at the bottom of the Layer Comps panel. The new comp reflects the current state of layers in the Layers panel.

3. Name the comp and choose options to apply to layers. Optionally, you can add descriptive comments.

To duplicate a comp:

- Drag the comp's thumbnail to the New Comps button.

VIDEO 6.3
Working with Layer Comps

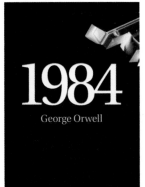

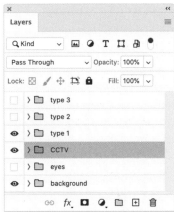

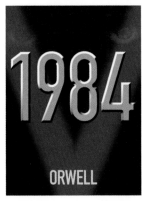

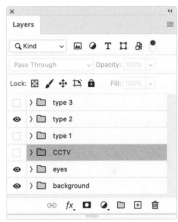

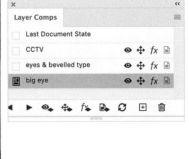
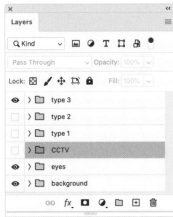

FIGURE 6.26 Using layer comps to create alternate versions of a book cover

To view layer comps:

Do one of the following:

- To view a layer comp, click the Apply Layer Comp icon next to a selected comp.
- To cycle through your layer comps, use the Previous and Next buttons at the bottom of the Layer Comps panel.

To restore the document to how it was before you chose a layer comp:

- Click the Apply Layer Comp icon next to Last Document State at the top of the Layer Comp panel.

If you change the configuration of a layer comp, you need to update it.

To update a layer comp:

- Click the Update Layer Comp button at the bottom of the panel.

You can export layer comps to individual files, as well.

To export layer comps:

1. Choose File > Export > Layer Comps To Files.

2. In the resulting dialog, choose the file type and set the destination.

Alternatively, you can export layer comps to PDFs.

To export layer comps to PDFs:

- Choose File > Export > Layer Comps To PDF.

In the resulting dialog, choose the destination and slideshow options.

Layer Masks and Vector Masks

Whatever the problem, masks are almost certainly part of the solution. You can use them for all kinds of creative and workaday Photoshop tasks, from isolating a subject from its background to creating composites by seamlessly blending images.

What makes masks so essential is that they allow you to work nondestructively. You can change your mind as often as you like without permanently messing up your image. Masking, rather than deleting parts of a layer, means that no pixels are harmed. By hiding parts of a layer you are revealing parts of the layer(s) below. If you go wrong, or just want to experiment with other solutions, you can restore the layer to the way it was.

"Mask it, don't delete it" are words to live by.

About Masks

There are two types of masks that you can apply to a layer:

- Layer masks are composed of gray-scale pixels that you edit with the painting or Selection tools (**FIGURE 7.1**).

- Vector masks are resolution independent and are created with the Pen or Shape tools. You cannot edit vector masks with the painting tools, but they are useful anytime you want sharp, defined edges (**FIGURE 7.2**).

For the majority of your masking needs, a layer mask is what's called for, because it uses pixels. They are best for organic subjects such as hair, fur, or anything where the hardness of the edge varies around the subject border. Layer masks are also the more flexible of the mask types, because they can be edited with Brush tools.

When it comes to man-made objects or subjects with defined outlines that are mainly straight lines and graceful curves, it may be preferable to use the close cousin of the layer mask: a vector mask. The typical workflow for these is to create a pen path around the subject and then convert the path to a vector mask.

Both layer and vector masks appear to the right of the layer thumbnail in the Layers panel.

VIDEO 7.1
Masks Overview

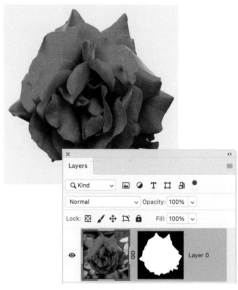

FIGURE 7.1 A layer with layer mask applied, which conceals the portion of the image outside the rose

FIGURE 7.2 An image with vector mask applied to isolate the road sign, revealing the color fill layer below

FIGURE 7.3 The "marching ants" indicating the active selection around the sunflower.

FIGURE 7.4 With the layer mask applied

FIGURE 7.5 The right side of the image is selected.

FIGURE 7.6 The layer mask conceals the right side of the image, revealing the layer below.

Adding Layer Masks

You can create a layer mask from a selection or by painting. Most often, you will make a mask from a selection, and then refine it with the painting tools. A "typical" mask looks like a stencil, with the white of the subject area surrounded by the black of the unselected areas.

To make a layer mask from a selection:

1. Make your initial selection (**FIGURE 7.3**).
2. On the Layers panel, click the image layer, layer group, or Smart Object to which you want to add the mask.
3. Do one of the following:
   - ▸ Click the Add Layer Mask button () at the bottom of the Layers panel.
   - ▸ Choose Layer > Layer Mask > Reveal Selection (**FIGURE 7.4**).

Depending on the image, you might want to hide rather than reveal the selected area.

To make a mask by hiding the selection:

1. Make your initial selection (**FIGURE 7.5**).
2. Alt/Option-click the Add Layer Mask button, or choose Layer > Layer Mask > Hide Selection (**FIGURE 7.6**). This is equivalent to choosing Select > Inverse before adding a layer mask.

TIP It's not uncommon to find that you have masked the opposite of what you intended. To rectify this, with the mask selected, press Ctrl/Command+I to invert the mask.

▶ VIDEO 7.2
Create and Edit a Layer Mask

Editing Layer Masks

You can continuously edit a layer mask to add to or subtract from the masked areas. Paint in black to hide; paint in white to reveal. The areas you paint in shades of gray appear in various levels of transparency—the lighter the gray the more of the layer is revealed and vice versa.

First, however, make sure you have the proper layer selected. When a layer mask thumbnail is selected, it has a small border that's easy to miss (**FIGURE 7.7**).

If in doubt, you can confirm you have a mask selected by looking in the Properties panel and the document title tab (**FIGURE 7.8**).

To edit a layer mask:

1. On the Layers panel, click the Mask thumbnail of the layer you want to edit. A border appears around the layer mask thumbnail. In addition, the foreground and background colors become grayscale when the layer mask is active.

2. Select any of the editing or painting tools.

3. Paint on the mask in black to hide portions of the layer, paint in white to reveal portions of the layer (**FIGURE 7.9**). To partially reveal the layer, paint the mask with a shade of gray—a lighter gray reveals more, a darker gray hides more.

> **TIP** Use a Foreground To Transparent gradient on the layer mask to create subtle transitions.

> **TIP** You can apply adjustments, such as Levels, and filters, such as Gaussian Blur, to layer masks to control their edges and density.

FIGURE 7.7 How do you know you have selected the layer mask? Look for the framing border around the corners of the thumbnail.

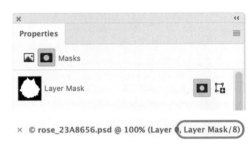

FIGURE 7.8 The Properties panel showing the layer mask and the document tab

FIGURE 7.9 Painting in black on the layer mask conceals more of the image—in this case the bottom of the sunflower stalk.

FIGURE 7.10 Add a layer mask to the texture layer.

FIGURE 7.11 Paint a "hole" through the texture on the layer mask to reveal the layer below. (In this case, we also changed the blending mode of the texture layer to Screen.)

FIGURE 7.12 Vary the size and hardness of your brush as well as its opacity.

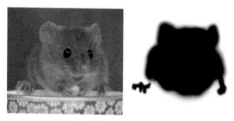

FIGURE 7.13 When painting the mask, toggle between black and white by pressing X. Toggle between viewing the mask and the image by Alt/Option-clicking the layer mask thumbnail.

TIP When painting on a layer mask, keep this in mind: "Black conceals, white reveals."

Painting a Layer Mask

Although a mask typically starts with a selection, it doesn't have to. Alternatively, you can paint the mask from scratch.

To paint a layer mask:

1. Select the layer, layer group, or Smart Object you want to mask.

2. With no selection active, Click the Add Layer Mask button (▣) (**FIGURE 7.10**).

3. With black as the foreground color, paint on the mask to conceal parts of the layer. Switch to white to paint back any areas that you want to restore (**FIGURE 7.11**).

4. Vary the brush size as necessary (**FIGURE 7.12**) with the [and] keys. Change the opacity of the brush with the number keys to paint in gray and partially reveal or conceal the layer.

To add to and subtract from a mask:

1. Reset the foreground and background colors to black and white by pressing D or clicking the small black/white swatches (▣).

2. Toggle between your foreground and background colors by pressing X or clicking the toggle switch (⇆) (**FIGURE 7.13**).

3. Paint on your mask layer with black to conceal an area, or touch up an area with white to reveal more.

4. If you get confused and paint in black when you should have painted in white or vice versa, just undo the step, press X to switch colors, and continue—layer masks are completely nondestructive.

Vector Masks

Vector masks are like layer masks, but have crisp, vector edges. If you have a manufactured subject requiring a crisp cut out, then a vector mask is the way to go. You can't vary the softness/hardness of the mask edge so that certain parts are harder or softer than others, nor can you paint on a vector mask with a brush or gradient. You can soften the edge uniformly using the Feather slider on the Properties panel.

If using the Pen or shape tools to make a vector mask, make sure to choose Path as the tool mode on the Options bar, otherwise you'll be making shape layers.

To add a vector mask:

1. Select the layer you want to mask.

2. Create a closed path with a Pen tool or Shape tool in Path mode (**FIGURE 7.14**).

3. Ctrl/Command-click the Add Mask icon at the bottom of the Layer panel. The first time you click creates a layer mask; the second time creates a vector mask. (A single click is enough if the layer already has a layer mask.) Alternatively, choose Layer > Vector Mask > Current Path.

To convert a custom shape to a vector mask:

1. With the Custom Shape tool, draw a shape layer (**FIGURE 7.15**). (If you are already in Path mode, skip to step 4.)

2. Select the shape layer with the Path Selection tool, and choose Edit > Cut (Ctrl/Command+X).

3. Select the layer that you want to mask, and choose Edit > Paste (Ctrl/Command+V). The shape will be pasted as a path to the layer (**FIGURE 7.16**).

FIGURE 7.14 Path mode on the Options bar

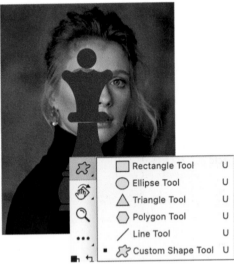

FIGURE 7.15 A vector shape drawn with the Custom Shape tool in Shape mode

FIGURE 7.16 The shape appears as unfilled vector outlines.

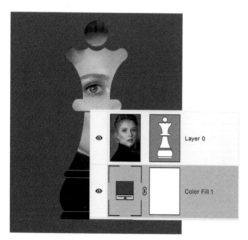

FIGURE 7.17 The result of the shape converted to a vector mask and how it appears on the Layers panel

FIGURE 7.18 Direct Selection tool (white arrow)

FIGURE 7.19 Editing the vector mask reveals the sky of the original image.

4. Ctrl/Command-click the Add Mask icon at the bottom of the Layers panel (or click the icon twice) to convert the shape to a vector mask.

5. (Optional) Unlink the vector mask and the layer, so that you can move the layer and vector mask independently to experiment with how the shape crops the image.

6. Add a solid color fill layer below (**FIGURE 7.17**).

> **TIP** If you prefer to start with a vector mask that shows or hides the entire layer, with nothing selected choose Layer > Vector Mask > Reveal All/Hide All.

To edit a vector mask:

1. On the Layers panel, click the Mask thumbnail of the layer you want to edit. If the layer has both a layer mask and a vector mask, select the vector mask, which is the second mask. A border appears around the vector mask thumbnail. In addition, a blue path line appears along the mask edge.

2. Choose the Direct Selection tool (A), which shares the same toolspace as the Path Selection tool (**FIGURE 7.18**).

3. Click the path edge to activate the path anchor points. Click and drag either from the anchor points or the path segments connecting the anchor points to adjust the shape of the vector mask (**FIGURE 7.19**).

VIDEO 7.3
Create and Edit a Vector Mask

About Masks and Alpha Channels

Layer masks, vector masks, alpha channels, quick masks, clipping masks—what's the difference? In one sense they are all the same: image overlays that show which parts of a layer are editable and which parts are masked, or protected from edits. Another thing to consider is that if you are working with any one of these things, you can easily convert it to any other. But what's most important is how they differ:

- An **alpha channel** (see Chapter 8) is a channel that represents a selection as a grayscale image (**FIGURE 7.20**). Alpha channels are independent of layers and color channels. You can convert alpha channels to and from selections or paths. Alpha channels do not, by themselves, change the appearance of your image, but give you the potential to do so. Some file formats, such as TIFF and PNG, use the term "alpha channel" to indicate which areas are transparent.

- A **quick mask** (see Chapter 8) is a temporary mask you create to restrict painting or other edits to a specific area of a layer (**FIGURE 7.21**). It's a selection in pixel form; instead of editing a selection marquee, you edit a quick mask using painting tools.

- A **layer mask** is an alpha channel that is attached to a specific layer (**FIGURE 7.22**). A layer mask controls which parts of a layer are revealed or hidden. Its thumbnail is next to the layer thumbnail in the Layers panel. A frame around the layer mask thumbnail indicates that it's selected.

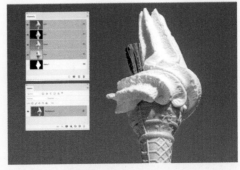

FIGURE 7.20 Don't let the name intimidate you: An alpha channel is just a saved selection.

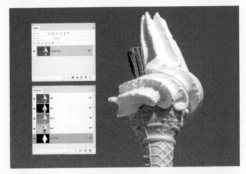

FIGURE 7.21 The masked portion of the image is shown in a color overlay (in this case, red). While in Quick Mask mode, an additional channel is shown on the Channels panel.

FIGURE 7.22 The only difference between an alpha channel and a layer mask is that the latter is attached to a layer.

FIGURE 7.23 A vector mask showing the pen path (selected with the Path Selection tool). Note that the masked areas show as gray on the vector mask thumbnail.

FIGURE 7.24 Making layer 1 into a clipping mask hides any parts of that layer that are outside the shape of the layer below.

- A **vector mask** is a mask made of resolution-independent vector paths, rather than pixels (**FIGURE 7.23**). Use a vector mask when having a precisely controlled mask edge is more important than being able to edit a mask with a brush. You create vector masks using the Pen or Shape tools.

- A **clipping mask** (see Chapter 9) is created when the content of one layer masks another layer or layers that are above it in the layer stacking order (**FIGURE 7.24**). Use a clipping mask to define the shape of one layer based on another, or to confine an adjustment to a specific layer, rather than have it affect everything below in the layer stack. The thumbnails of a clipped layer are indented, with an arrow pointing to the layer below. The name of the clipped base layer is underlined.

Creating a Simple Composition with Masks

Here's a simple twist: Masks don't have to be made from the layer to which they are attached.

To use selections from one layer as a layer mask for another:

1. Make a selection on one layer, even a rough one with the Quick Selection tool, as in **FIGURE 7.25**.

2. In the Layers panel, select a different layer to mask, and click the Add Layer Mask button (**FIGURE 7.26**).

3. Double-click the Layer Mask icon to bring up the Properties panel to fine-tune the Density and Feather settings as needed (**FIGURE 7.27**).

FIGURE 7.25 Target the waterfall layer, and use the Quick Selection tool to make a rough selection of the waterfall, then use the Inverse command (Select > Inverse) so that *everything but* the waterfall is selected.

FIGURE 7.26 Turn on the visibility of the frame layer, then select the layer and click Add Layer Mask.

FIGURE 7.27 To make the water translucent and to soften the edge of the mask, reduce the Density and increase the Feather settings on the Properties panel.

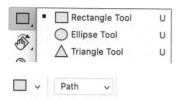

FIGURE 7.28 Use the Rectangle tool in Path mode to create the vector mask shape.

FIGURE 7.29 So that the interior rectangle knocks a hole through the larger rectangle, choose Exclude Overlapping Shapes.

FIGURE 7.30 Draw two rectangles: one outside and one inside the frame.

4. Use a vector mask to hide the white areas around and inside the frame. To create a vector mask, choose the Rectangle shape tool and change the tool mode to Path (**FIGURE 7.28**).

5. From the Path Operations menu on the Options bar, choose Exclude Overlapping Shapes (**FIGURE 7.29**).

6. Draw an exterior rectangle around the frame. Start with a rectangle that is close in size, then from its corner handles adjust the size and rotation of the rectangle to match the frame. Repeat this to draw an interior rectangle (**FIGURE 7.30**).

7. Hold Ctrl/Command and click Add Mask to convert the vector path into a vector mask (**FIGURE 7.31**).

TIP You can drag to move a layer mask from one layer to another. If you hold Alt/Option you will duplicate the mask. If there's already a mask on that layer, you'll be asked if you want to replace it.

VIDEO 7.4
Create a Composition with Layer Masks and Vector Masks

FIGURE 7.31 The finished result and the layer mask and vector mask shown on the Layers panel.

Layer Masks and Adjustment Layers

Every time you add an adjustment layer, (see Chapter 9) you get a blank layer mask thrown in. If you have an active selection when you choose the adjustment layer, the non-selected areas will be black on the resulting layer mask. If you add the adjustment layer without an active selection, the layer mask attached to it will start out white, and so it has no effect—until you paint on it in black or gray to limit the adjustment to specific parts of the image.

To use a layer mask with an adjustment layer:

1. Make a selection of the part of the image you want to adjust, and choose Add Layer Mask (**FIGURE 7.32**). Depending on the result you're after, you may want to inverse the selection (Ctrl+Shift+I/Command+Shift+I).

2. Choose an adjustment layer from the menu at the bottom of the Layers panel (●.).

3. If necessary, choose a painting tool and paint on the layer mask in black or white to refine the mask.

> **TIP** If the adjustment is intended to affect less than half of the image, fill the layer mask with black, and then paint in the adjustment by painting with white on the layer mask.

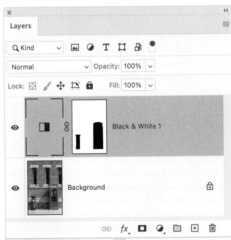

FIGURE 7.32 The Object Selection tool is used to select the mailbox and telephone box. The selection is inversed (Select > Inverse), and then a black and white adjustment layer is chosen (Layer > New Adjustment Layer > Black & White), turning the background monochrome.

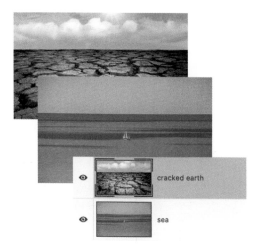

FIGURE 7.33 In this example, the document is made up of two image layers.

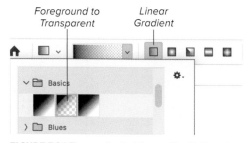

Foreground to Transparent   *Linear Gradient*

FIGURE 7.34 The gradient picker on the Options bar

FIGURE 7.35 Build up the gradient layer mask with multiple swipes to achieve the desired result.

Gradient Layer Masks

You can use a gradient layer mask to achieve a subtle transition from one image to another.

To apply a gradient mask:

1. Create a document with two or more image layers (**FIGURE 7.33**).

2. Add a layer mask to the top layer by selecting the layer and clicking the Add Layer Mask icon ().

3. If necessary, restore the foreground colors to black and white by pressing X.

4. Select the Gradient tool. From the gradient picker, choose the Foreground To Transparent gradient and a linear gradient (**FIGURE 7.34**).

5. Making sure the layer mask is selected, make multiple swipes of the Gradient tool (from right to left in this case) to build up the gradient layer mask cumulatively (**FIGURE 7.35**).

> ▶ **VIDEO 7.5**
> **Working with Gradient Layer Masks**

You can also use a gradient layer mask in combination with adjustment layers to achieve a seamless transition from an adjusted to a non-adjusted portion of the image. This is particularly useful when you want to fix exposure in an image by darkening or lightening one side of the image, leaving the other side unchanged.

To apply a gradient mask to an adjustment layer:

1. Add a Curves adjustment layer (**FIGURE 7.36**).

2. Pull up the curve from the shadow area to increase the exposure (**FIGURE 7.37**).

3. To restore part of the image (in this case, the sky) to how it was before the adjustment, use a Foreground (black) To Transparent gradient on the layer mask that comes with the Curves adjustment: Select the layer mask, and drag from the top about halfway down the image. If necessary, you can make multiple swipes with the Gradient tool, to build the gradient layer mask cumulatively (**FIGURE 7.38**).

FIGURE 7.36 The image is intentionally underexposed to enhance the intensity of the sunrise.

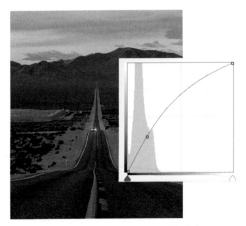

FIGURE 7.37 Pulling up the curve adds light, improving the exposure in the lower two thirds of the image, but at the same time causes the sky to become overexposed.

FIGURE 7.38 To restore the sky, a Foreground (black) to Transparent gradient is added to the layer mask, dragging from the top of the image. Where the gradient is black it protects the layer from the curve adjustment.

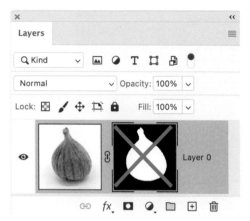

FIGURE 7.39 The layer mask disabled

FIGURE 7.40 Different ways of viewing the same thing: the image with layer mask applied (left); viewing the mask itself

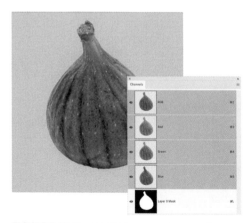

FIGURE 7.41 Viewing the layer mask as a color overlay

TIP When viewing a mask as a color overlay, the mask color has no effect on how the masked areas of the layer are protected.

Managing Layer Masks

Here are some essential techniques for keeping on top of your layer masks.

While editing, it's useful to swap back and forth between seeing a layer with and without its mask. Likewise, it's sometimes easier to understand and edit a layer mask by viewing it in grayscale or as a color overlay on the layer. You can toggle these views as often as necessary.

To disable or enable a layer mask:

1. To see the layer without its mask, Shift-click the mask thumbnail in the Layers panel. A red X appears over the mask thumbnail, and all the layer's content becomes visible (**FIGURE 7.39**).

2. Shift-click again to enable the mask.

To view a mask as grayscale:

1. Alt/Option-click the layer mask thumbnail to view just the grayscale mask (**FIGURE 7.40**).

2. Alt/Option-click the layer mask thumbnail again to view the layer. Alternatively, click the eye icon on the Properties panel.

To view the mask as a color overlay:

1. Alt/Option+Shift-click the layer mask thumbnail to view the mask as a color overlay. Or, open the Channels panel and click the eyeball next to the layer mask (**FIGURE 7.41**).

2. If you need to change the layer mask color—perhaps for more contrast against the colors in the image—double-click the layer mask channel in the Channels panel and choose a new mask color in the Layer Mask Display Options dialog.

To apply or delete a layer mask:

1. To permanently apply a layer mask, right-click the layer mask thumbnail, then choose Apply Layer Mask. But, be careful: any pixels made fully or partially transparent by the layer mask are permanently deleted, and can't be recovered except through Undo.

2. To delete a layer mask without applying the changes, right-click the layer mask thumbnail and choose Delete Layer Mask (**FIGURE 7.42**).

FIGURE 7.42 Deleting the layer mask restores the layer to its unmasked state.

Mask Properties

The Properties panel (**FIGURE 7.43**) is the place to adjust the density (opacity), the amount of feather applied to the edges of a selected layer mask or vector mask, and more.

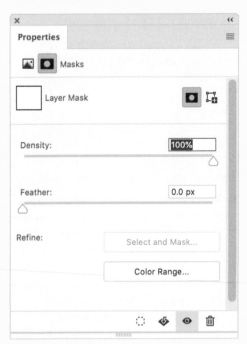

FIGURE 7.43 Double-click a layer mask to bring up the Properties panel.

- **Density** lets you control mask opacity non-destructively. At 100%, the mask is opaque and conceals any underlying area of the layer. Reducing the density shows more of the area under the mask.

- **Feathering** blurs the edges of the mask to create a softer transition between the masked and unmasked areas. The feathering goes both ways from the mask edge.

You can also apply a couple of Refine options to layer masks.

- **Select and Mask** lets you modify mask edges using the Select and Mask dialog box, as well as view the mask against different backgrounds. By default, changes are output to a new mask.

- **Color Range** lets you refine the mask using the color range options. The color range you create will intersect with already selected portions of the mask, meaning that you'll end up with less of the layer visible than before.

The Invert button reverses white and black. Hidden areas become visible, and vice versa.

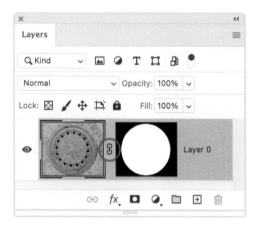

FIGURE 7.44 The link between layer and mask

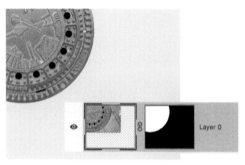

FIGURE 7.45 When a layer and its mask are linked, both are moved or transformed together.

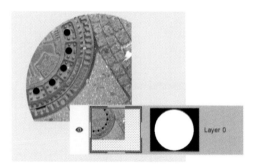

FIGURE 7.46 When they are not linked, either the mask or the layer can be moved independently.

Linked layers and masks

By default, a layer or group is linked to its layer mask or vector mask—you'll see a link icon between the thumbnails in the Layers panel (**FIGURE 7.44**). When you move or transform (scale, rotate, etc.) either the layer or the mask, both layer and mask are moved or transformed together (**FIGURE 7.45**). Unlinking the layer from its mask lets you move the mask's boundaries separately from the layer (**FIGURE 7.46**).

To unlink layers and masks:

- Click the link icon to link/unlink the layer and its mask.

Selections from masks

Sometimes you'll want to activate a selection from an existing mask. This is easy since you've already done the work to make the mask in the first place.

To create a selection from a mask:

Do one of the following:

- If the layer with the mask is selected, choose Select > Load Selection. The mask will appear in the list of channels from which a selection can be made. The Load Selection dialog also offers you the ability to invert the selection or add, subtract, or intersect it with an existing selection (**FIGURE 7.47**).

- Hold Ctrl/Command and click the mask's thumbnail for a marching ants selection.

- To add the selection to an existing selection, hold Shift and Ctrl/Command-click the mask's thumbnail.

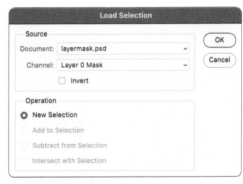

FIGURE 7.47 The Load Selection dialog

- To subtract from an existing selection, hold Alt/Option and Ctrl/Command-click the mask's thumbnail.

- To intersect with an existing selection hold Alt/Option+Shift and Ctrl/Command-click the mask's thumbnail.

Layer Mask Shortcuts

When working with layer masks you can change several variables: foreground color, brush size, brush hardness, and opacity. Get familiar with the shortcuts and you will be able to work much faster and easier.

- Disable/Enable a mask: Shift-click the mask thumbnail.

- View just the mask: Alt/Option-click the mask thumbnail.

- View the mask as color overlay: Alt+Shift-click/Option+Shift-click the mask thumbnail.

- Toggle Foreground/Background color: X

- Restore Foreground/Background color to default black and white: D. (If a layer mask is selected, the default foreground/background colors are reversed.)

- Change Brush Size: [(Left Bracket key) to go smaller;] (Right Bracket key) to go bigger.

- To change Brush Hardness: Shift+[to go softer in 25% increments; Shift+] to go harder in 25% increments.

- Painting with an opacity of less than 100% on a layer mask is a great way to achieve transparency effects. The higher the brush's opacity the more it reveals or conceals. To change brush opacity, use your keypad numbers: 1 for 10%, 5 for 50%, 0 for 100%, and so on.

- You can access the layer mask of a selected layer by pressing Ctrl/Command+\.

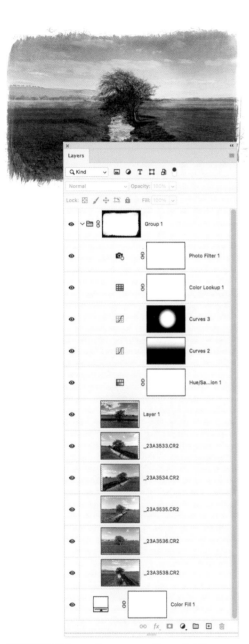

Masking groups

Masks can be applied to layer groups as well as to individual layers. If you want several layers to be affected by the same mask, rather than copy the same layer mask to multiple layers, put the layers in a group and mask the group.

To mask a group of layers:

1. Select the layers you want to mask and click the Group icon (▢) at the bottom of the Layers panel to add them to a group.

2. With the Group icon selected, make a selection to use as a mask, or activate a saved selection.

3. Click Add Layer Mask (**FIGURE 7.48**).

It's also possible for each layer in the group to have its own mask. This is how you get around the limitation that a layer can have only one layer mask and one vector mask. This works even if there is only one layer in the group.

FIGURE 7.48 Several exposures (from slightly differing perspectives) of the subject are blended with a combination of layer blending modes and adjustment layers. The combined result is then masked by the rough-edged frame, which is applied as a layer mask to the group.

Masks on Smart Objects

There are pros and cons to applying layer masks to the contents of a Smart Object, rather than to the Smart Object itself. A mask applied to a Smart Object isn't protected from destructive editing (**FIGURE 7.49**) the way the contents of the Smart Object are. If you apply repeated transformations, the mask may be degraded. On the other hand, when you apply the mask to the Smart Object, rather than to its content, the mask is visible in the Layers panel, making this approach a more flexible way of working (**FIGURE 7.50**). This is especially true if the Smart Object contains multiple layers; a mask on the Smart Object masks everything in it. (For more on Smart Objects, see Chapter 14.)

If you want to make a Smart Object of a layer with a layer mask applied, but want to continue working on that layer mask, do the following.

To add a mask to a Smart Object:

1. Activate the selection from the layer mask by Ctrl/Command-clicking the layer mask thumbnail.

2. Right-click the layer mask thumbnail and choose Delete Layer Mask to delete the layer mask without applying it.

3. Right-click the layer thumbnail and choose Convert To Smart Object.

4. Convert the active selection into a layer mask for the Smart Object by clicking the Add Mask icon.

> **TIP** Unlike with a regular image layer, you can't permanently apply a layer mask to a Smart Object when deleting the layer mask.

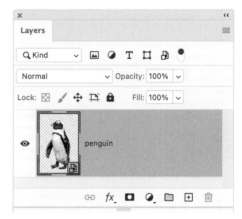

FIGURE 7.49 If you add the mask first, and then convert to a Smart Object, then the layer mask will not be visible or easily editable.

FIGURE 7.50 Convert the layer to a Smart Object first, then add the layer mask to give yourself more options when editing the file.

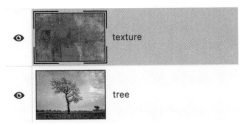

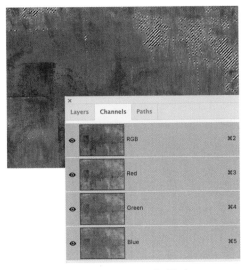

FIGURE 7.51 The composition is comprised of two layers.

FIGURE 7.52 The Channels panel with the composite (RGB) channel at the top of the list

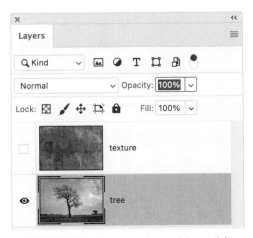

FIGURE 7.53 Hide the texture layer while retaining the selection made from RGB channel.

Luminosity Masks

A luminosity selection is proportional to the luminosity in the image. The lighter an area is, the more selected it will be. When you create a layer mask from such a selection, it will be a grayscale version of the image, "hiding" the darker parts and revealing the lighter parts. Amongst other things, this is useful for combining texture with an image.

To make a luminosity mask:

1. Create a document of two of more layers.

2. Select the top layer—in this case the texture—from which you will create the mask (**FIGURE 7.51**).

3. Go to the Channels panel, and Ctrl/ Command-click the RGB composite channel to make the selection of the gray values of the image. You will see a marching ants selection on the image (**FIGURE 7.52**).

4. Return to the Layers panel, hide visibility of the texture layer, and select the layer beneath (**FIGURE 7.53**).

5. With the selection still active, click Add Layer Mask (**FIGURE 7.54**).

6. Create a solid color fill layer at the bottom of the layer stack (**FIGURE 7.55**).

7. (Optional) Select the Layer Mask and press Ctrl/Command+L to bring up the Levels dialog. Move the midpoint slider left to reduce the contrast (revealing more of the image) or right to increase the contrast (concealing more of the image) (**FIGURE 7.56**).

FIGURE 7.54 When you first add the layer mask, it may fade (mask) the image too much.

▶ **VIDEO 7.6**
Create a Luminance Mask

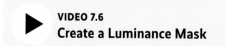
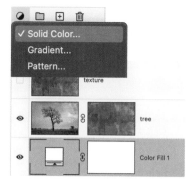

FIGURE 7.55 The solid color (in this case white) fill layer below will give the image more presence—for different effects experiment with using different colors.

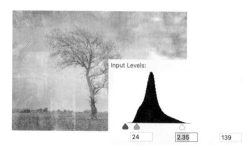

FIGURE 7.56 Adjust the contrast of the layer mask using Levels.

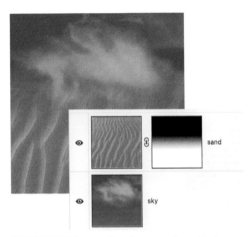

FIGURE 7.57 The gradient layer mask applied to the top layer (sand) reveals part of the layer below (sky).

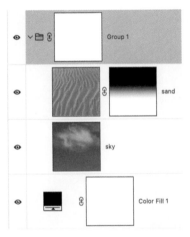

FIGURE 7.58 Apply the mask to the group.

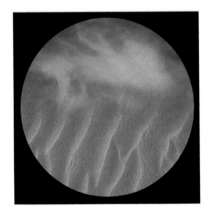

Combining Layer Masks and Vector Masks

If you have multiple layers with layer masks applied to them, you can mask the cumulative effect of those layers with a vector mask.

To mask more than one layer with a vector mask:

1. Create a document of two layers, and apply a gradient layer mask to the top layer, revealing part of the layer below (**FIGURE 7.57**).

2. Select the layers and press Ctrl/ Command+G to put them in a group.

3. Click the newly created group to make it active, and then Ctrl/Command-click the Add Mask icon (**FIGURE 7.58**).

4. With the Ellipse tool drawing in Path mode, draw a circle on the vector mask (**FIGURE 7.59**).

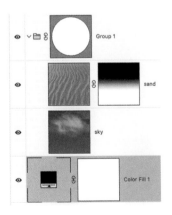

FIGURE 7.59 Draw with the Ellipse tool on the vector mask to mask the contents of the group, revealing the color fill layer beneath.

It's not often that you need to do so, but you can have both a layer mask and a vector mask attached to a single layer. (Just click the Layer Mask icon twice.)

Or, if you want to add different qualities of mask edge to different parts of the image, you can copy the layer, adding a layer mask to one copy and a vector mask to the other.

To apply a vector and a layer mask to copies of the same layer:

1. Add a vector mask (in **FIGURE 7.60**, a circle) to the layer.

2. Press Ctrl/Command+J to duplicate the layer (**FIGURE 7.61**).

3. Delete the vector mask from the top layer, and replace it with a layer mask (around the shape of the leaf) (**FIGURE 7.62**).

4. Adjust the feathering of the layer mask as necessary using the Properties panel.

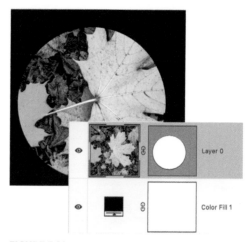

FIGURE 7.60 The circle vector mask reveals the color fill layer beneath.

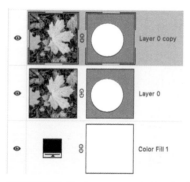

FIGURE 7.61 The Layers panel after duplicating the leaf layer

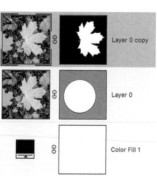

FIGURE 7.62 The layer mask provides a softer, more organic edge for the leaf, while the vector mask provides a sharp-edged enclosing circle.

Further Selection Techniques

Some selections are fast and easy, so fast and so easy they seem like magic; others take time and patience. Sometimes all you need is a rough selection; sometimes you need a painstakingly accurate selection. It all depends upon the properties of the image and what you're trying to achieve. This chapter explores more advanced selection techniques, such as how to use Color Range and the Select And Mask workspace to create—and save—more nuanced selections. You'll see how to automatically identify and select the subject of an image or distinct objects in an image, as well as take a look at the wizardry of Sky Replacement and how to select the in-focus area of an image.

Selecting a Color Range

Using Color Range you can select a specified color or range of colors within an existing selection or within an entire image. Color Range allows you to partially select pixels—just like painting on a mask with gray.

To select a color range:

1. Choose Select > Color Range.

2. From the Select menu, choose Sampled Colors.

3. Click with the Eyedropper to sample the colors you want included in the selection. Each click makes a new selection, indicated in white and grays in the preview area (**FIGURE 8.1**).

4. Use any of the following options to refine the selection:

   ▸ To add colors, hold Shift (or select the Plus Eyedropper), and click.

   ▸ To remove colors, Alt/Option-click (or select the Minus Eyedropper), and click.

   ▸ Adjust the Fuzziness slider: This is similar to the Magic Wand's Tolerance setting. A low value restricts the color range; a higher value increases it.

   ▸ Turn on Localized Color Clusters to determine how near or how far a color must be from the sample points to be included in the selection. Reduce the range to exclude colors that are further from the selected colors.

   ▸ Check Invert to inverse the resulting selection.

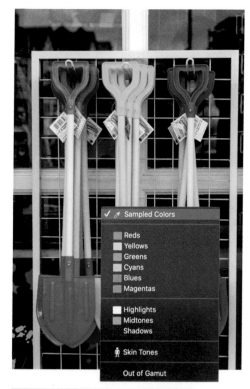

FIGURE 8.1 Making a Color Range selection by sampling colors from the image

FIGURE 8.2 Selecting skin tones with Color Range

▶ Choose a Selection Preview to determine how the selection you are making is displayed on the image itself. We prefer None, but you have the option of Grayscale, Black Matte, White Matte, and Quick Mask.

With Color Range you can also select skin tones and detect faces. If you want to make an adjustment that preserves skin tones while you adjust the color of everything else in the image, select Invert below the eyedropper samplers.

To select skin tones:

1. Choose Select > Color Range.

2. From the Select menu, choose Skin Tones to select colors that resemble skin tones (**FIGURE 8.2**).

3. Select Detect Faces and then adjust the Fuzziness slider.

TIP To change the preview, choose a display option: Selection for a preview of your selection in white (selected areas), black (unselected areas), and gray (partially selected areas), or Image to preview the entire image.

TIP Press Ctrl/Command to toggle between Image preview and Selection preview in the Color Range dialog.

TIP If you see the message "No pixels are more than 50% selected," the marching ants will not be visible when you return to the image.

Using Select Subject

Select Subject enables you to select the most prominent subject in an image with a single click. Although it is unlikely to get you exactly what you're after, it is a good starting point (**FIGURE 8.3**). You can then refine your selection using other tools.

To use Select Subject:

Do one of the following:

- Choose the Object Selection, Quick Selection, or Magic Wand tool, then click Select Subject on the Options bar.
- Choose Select > Subject.
- In the Select And Mask workspace, click Select Subject on the Options bar.

> ▶ **VIDEO 8.1**
> **Making Color Range Selections**

FIGURE 8.3 Select Subject incorrectly assumes we want the leaf at the top included in the selection, but we can easily subtract it from the active selection to refine the results.

Using the Object Selection Tool

If the objects in your image are clearly defined, a single click may be all you need to select them. Found in the Tools panel and the Select And Mask workspace, the Object Selection tool shares the same tool space as the Magic Wand and the Quick Selection tool. It's useful for selecting a single object or part of an object in an image. When Object Finder is checked on the Options bar, the Object Selection tool automatically detects objects in the image (**FIGURE 8.4**). Move the pointer over a detected object and you'll see a semi-transparent overlay (blue by default).

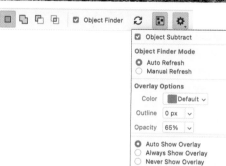

FIGURE 8.4 The Object Finder options control the behavior of the Object Finder tool

FIGURE 8.5 Choose a mode for the Object Selection tool on the Options bar.

FIGURE 8.6 Hover over an object and click to select it.

Click the object you want to select. If the Show All Objects icon (⊞) is selected on the Options bar, all identifiable objects are covered by the semitransparent overlay.

To select an object with the Object Selection tool:

Do one of the following:

- Draw a lasso or rectangle to surround the object (**FIGURE 8.5**). The Object Selection tool selects the object inside the defined region.

- Hover over a well-defined object until it is indicated with a color overlay and click (**FIGURE 8.6**).

Select Subject versus Object Selection

What's the difference between Select Subject and the Object Selection tool? Select Subject selects all the main subjects in the image, while the Object Selection tool lets you select one of the objects or part of an object in an image that has multiple objects (**FIGURE 8.7**).

FIGURE 8.7 Select Subject selects all three pine cones (note the marching ants). To make a selection of just one, drag a marquee with the Object Selection tool.

To create masks for selected objects:

- To create masks for all identifiable objects in the image choose Mask All Objects from the Layer menu or the Layers panel menu. Separate layer masks attached to a Layer group are created for each object (**FIGURE 8.8**).

> **TIP** Press and hold N to toggle the Show All Objects option on and off.

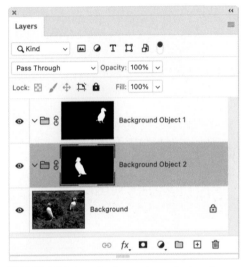

FIGURE 8.8 Automatically create masks for all identified objects in the image.

Working with Select And Mask

The Select And Mask workspace is a powerful, dedicated workspace for making precise selections and masks. Whether you prefer to make a rough selection first—before going to Select And Mask—or you prefer to start from scratch in the Select And Mask workspace is up to you.

To use the Select And Mask workspace:

1. Use Select Subject, the Object Selection tool, or the basic selection tools to make a rough selection.

2. Click Select And Mask to enter the Select And Mask workspace. (Or, you can skip step 1 and make the selection entirely within Select And Mask.)

3. Make any necessary adjustments using the Select And Mask tools (**FIGURE 8.9**). These work the same way as in the Photoshop workspace: Click or click-drag with the Quick Selection tool, paint in Add or Subtract mode with the Brush tool, and adjust the border area with the Refine Edge Brush tool (useful for subtle areas and fine detail).

FIGURE 8.9 The Select And Mask tools. The Quick Selection, Brush, Object Selection, and Lasso tools are "hard-edged" tools. Use these on areas that you want to be fully opaque or fully transparent. For areas with semitransparent or transitional details, use Refine Edge Brush.

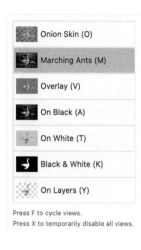

Press F to cycle views.
Press X to temporarily disable all views.

FIGURE 8.10
Different ways of viewing the same information: what's selected and what's not selected.

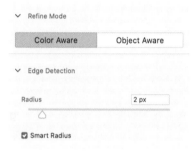

FIGURE 8.11 Determine the selection border's size.

FIGURE 8.12 Fine-tune the selection edge with the global refinements sliders. Use the sliders in conjunction with each other. In general smaller numbers give better results.

4. Adjust the View and Refine mode settings in the Properties panel to help you fine-tune your selection (**FIGURE 8.10**). Refine mode lets you work in Color Aware (best for simple subjects against contrasting color backgrounds) or Object Aware (best for more complex selections) mode (**FIGURE 8.11**).

5. Choose the Edge Detection settings: Radius is suited to subjects with uniformly edged borders; choose a small radius for sharp edges, a larger one for softer edges. Smart Radius varies the width of the border around the selection and is helpful for non-uniform edges such as in portraits, where the hair might have a different quality of edge to the shoulders.

6. Adjust the Global Refinements sliders to adjust the selection edges (**FIGURE 8.12**): Smooth (reduces bumps), Feather (blurs), Contrast (sharpens), or Shift Edge (moves the border inside or outside the selection, effectively contracting or expanding the selection).

7. Choose an Output To setting for your selection (**FIGURE 8.13**). When turned on, the Decontaminate Colors option replaces color fringing with the color of nearby pixel. Because it changes the color of the edge pixels, it requires output to a new layer or document.

8. Click OK to leave Select And Mask and return to the main Photoshop interface.

TIP Press F to cycle through the View modes; press X to temporarily disable them.

TIP With **Show Edge** turned on (and Show Original turned off) for Edge Detection you see the width of the edge that the Radius setting affects. You also see the edge modifications produced by the Refine Edge Brush tool and Refine Hair button.

TIP Although the global Feather and Contrast settings seemingly work in opposition, it's common to apply both to an edge.

TIP Slightly shifting a border inward with Shift Edge can help remove fringing on selection edges.

To mask a subject with flyaway hair:

1. In the Select And Mask workspace click Select Subject.

2. Choose a View mode that will show the problems with the selection edge. The "best" mode depends upon the colors in your image; for the example, we chose White.

3. Choose Object Aware as the Refine mode.

4. Click Refine Hair to exclude any negative space surrounded by hair. If further refinements are needed, paint over those hairs with the Refine Edge Brush tool (**FIGURE 8.14**).

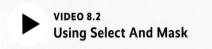

VIDEO 8.2
Using Select And Mask

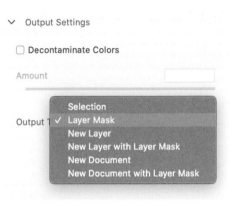

FIGURE 8.13 Choose what happens to your selection when you exit Select And Mask.

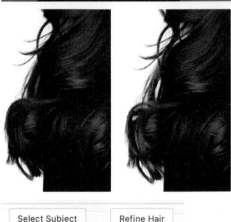

FIGURE 8.14 Select And Mask makes masking hair easy.

Replacing Skies

If you returned from a photo shoot and the sky in your images is a bit flat or washed out, you can add drama with Sky Replacement.

1. Choose Edit > Sky Replacement.

2. Choose a new sky from the presets or add an image of your own. Photoshop selects and masks the sky of the original image to reveal the new sky in its place (**FIGURE 8.15**).

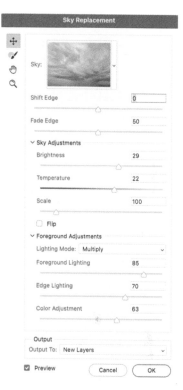

FIGURE 8.15 Replacing a washed out sky and the resulting layers

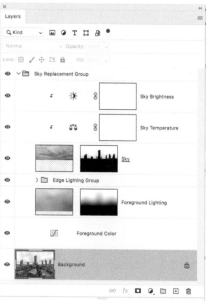

3. (Optional) Adjust Shift Edge to move the border between the sky and the original photo or Fade Edge to tailor the amount of feathering along that edge.

4. (Optional) To further refine the sky, adjust the Brightness, Temperature, Lighting mode (blending mode used), Lighting Adjustment (exposure of the original image at the blended edge), or Color Adjustment (how the foreground is matched with the sky colors).

5. Choose your preferred Output setting (New Layers creates a layers group named Sky Replacement Group and Duplicate layer outputs a single flattened layer), and click OK.

TIP Use the Scale option to resize the sky image and select Flip to flip it horizontally.

VIDEO 8.3
Replacing a Sky

Add Your Own Skies

Just because we live under the same sky, we don't want all our skies to look the same. Photoshop gives you options:

- **Import Skies From Images** lets you create new sky presets from your own images.
- **Import Skies From Sky Presets** lets you import sky presets from SKY files.
- **Get More Skies** takes you to the Adobe Discover website where you can download more free images or presets (**FIGURE 8.16**).

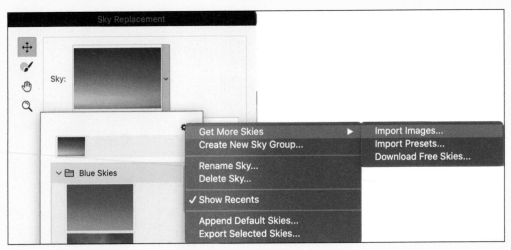

FIGURE 8.16 There's no limit to the skies you use.

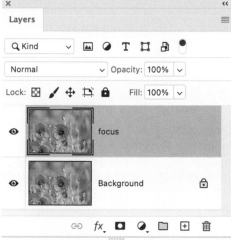

FIGURE 8.17 Duplicate the layer and name the copy.

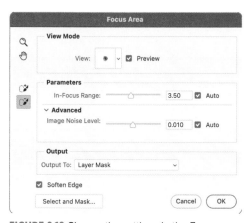

FIGURE 8.18 Choose the settings in the Focus Area dialog.

Using Select Focus Area

Select Focus Area lets you select the in-focus parts of an image. For instance, you could select the focused parts of an image, then invert the selection and defocus the rest of the image to draw attention to focused areas.

To use Select Focus Area:

1. To make the result nondestructive, duplicate the layer you are working on (Ctrl/Command+J) and name the copy (**FIGURE 8.17**).

2. On the duplicate layer, choose Select > Focus Area.

3. Refine the selection with the Focus Area Add and Focus Area Subtract tools and by adjusting the In-Focus Range slider. With the slider at 0, the whole image is selected. Move the slider to the right, and only the focused parts of the image remain selected.

4. (Optional) Use the Advanced > Image Noise Level slider to control any noise in the image.

5. If you want to feather the edges of the selection, select Soften Edge. You can also click Select And Mask to further refine the selection.

6. Choose an appropriate output option, which are the same as those in the Select And Mask workspace. We chose Layer Mask (**FIGURE 8.18**).

7. Return to the background layer, and choose Filter > Blur > Lens Blur and adjust the settings to your liking (**FIGURE 8.19**).

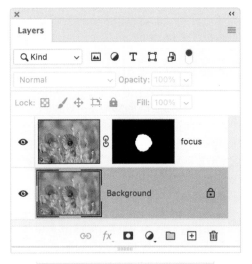

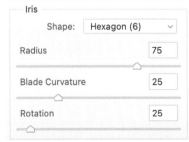

FIGURE 8.19 Apply blurring to the background layer.

Using Quick Masks

A Quick Mask is somewhere between a selection and an alpha channel. Like a selection, a Quick Mask goes away when not in use; like an alpha channel, a transparent color overlay represents the masked areas of the image. A Quick Mask channel shows up in the Channels panel—but only when it's active. And you can edit the shape of the Quick Mask with painting tools or filters. When making complicated selections, you can be more accurate using painting tools on a Quick Mask than with the basic selection tools.

The thing about Quick Mask is that it's, well, quick—and temporary. You convert a selection into a Quick Mask to refine it with painting tools, and then convert it back to a selection when you're done. You can then save it as an alpha channel or a layer mask. Quick Masks don't offer options that you can't find elsewhere, but some people find them more convenient.

Start with a basic selection, and then refine it in Quick Mask mode. If you prefer, you can create the mask entirely in Quick Mask mode. A color overlay differentiates the protected and unprotected areas. When you leave Quick Mask mode, the color overlay becomes an active selection (marching ants).

FIGURE 8.20
Creating a
Quick Mask

To apply a Quick Mask:

1. Make a selection using any selection tool.

2. Click the Quick Mask Mode button (Q) at the bottom of the Tools panel (**FIGURE 8.20**).

3. Select a painting tool. The foreground and background colors automatically become black and white.

4. Paint in white to remove from the color overlay and add to the selected parts of the image. Paint in black to add to the color overlay and deselect areas. Paint in gray or another color to create a semitransparent area, useful for softening edges.

5. Click the Standard Mode button or press Q to return to your original image. A marching ants selection now surrounds the unprotected area of the Quick Mask.

Quick Mask Tips

By default, Quick Mask mode uses a 50% red overlay to indicate the protected area. If you need to make the mask more visible against the colors in the image you can change both color and percentage by double-clicking the Quick Mode button (**FIGURE 8.21**).

To toggle between whether the color indicates the selected or masked areas, Alt/Option-click the Quick Mask Mode button. These settings affect only the appearance of the mask and not how underlying areas are protected.

Although semitransparent areas may not appear selected when you exit Quick Mask mode, they actually are. A marching ants selection indicates the transition between pixels that are less than 50% selected and those that are more than 50% selected.

FIGURE 8.21 Choose how the Quick Mask will appear.

A temporary Quick Mask channel appears in the Channels panel while you're in Quick Mask mode—and disappears when you return to standard mode. If you want to convert this temporary mask to a saved alpha channel, switch to standard mode and choose Select > Save Selection.

Saving Selections (Alpha Channels)

Having invested time in making a selection, you may want to save it so that you can recall it later. You can save any selection as an alpha channel. Alpha channels are saved with the document as potential selections, which you can load on an as-needed basis. A typical alpha channel will look like a stencil—the white areas indicating what's selected and the black areas indicating what's unselected or masked.

Because alpha channels are grayscale images, you can edit them like any other image with painting tools, editing tools, and filters. Painting in gray creates a semi-transparent area.

To save a selection to a new channel:

1. Use any Selection tool to make a selection

2. On the Channels panel, click Save Selection As Channel at the bottom of the panel. This will create a new alpha channel named Alpha 1, Alpha 2, and so on. You can double-click the channel name to rename it (**FIGURE 8.22**).

FIGURE 8.22 Click the Save Selection icon to convert an active selection into a saved alpha channel.

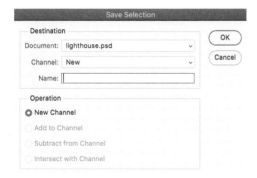

FIGURE 8.23 For more options, use the Save Selection dialog.

To save a selection to an existing channel:

1. Choose Select > Save Selection.

2. In the Save Selection dialog (**FIGURE 8.23**), choose a destination image for the selection from the Document menu. By default, the selection is saved in your active image. Most of the time this is what you'll want, but you can also save the selection as a channel in another image—so long as it is open and has the same pixel dimensions.

3. Specify a destination channel for the selection from the Channel menu: Choose to save the selection to an existing channel in the selected image or to a layer mask if the image contains layers. When saving the selection to an existing channel, choose whether you want to replace the existing channel, add to it, subtract from it, or intersect with it.

To create an alpha channel from scratch:

1. Click the New Channel button at the bottom of the Channels panel.

2. Paint on the new channel to mask the image.

TIP Alt/Option-click the New Channel button at the bottom of the Channels panel to change the color and opacity of the mask and whether the color indicates the masked or selected areas. To change these options for an existing channel, select Channel Options from the Channels panel menu or double-click the channel's thumbnail in the Channels panel.

To load a saved selection:

Do one of the following:

- Drag the saved selection (alpha channel) onto the Load Selection button (⬚).

- Ctrl/Command-click the channel containing the selection you want to load.

- Select the alpha channel, click Load Selection at the bottom of the Channels panel, then click the composite color channel.

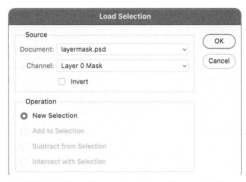

FIGURE 8.24 Loading a selection with a selection already active

To combine selections:

Do one of the following:

- Ctrl/Command+Shift-click the channel to add to the existing selection.

- Ctrl+Alt-click/Command+Option-click the channel to subtract from the existing selection.

- Ctrl+Alt+Shift-click/Command+Option+Shift-click the channel thumbnail to load the intersection of the saved selection and the existing selection.

- Choose Select > Load Selection and specify the Source options in the Load Selection dialog. If the image already has a selection active, choose how to combine the selections (**FIGURE 8.24**).

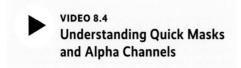

VIDEO 8.4
Understanding Quick Masks and Alpha Channels

Adjustment Layers and Image Adjustments

Using adjustments layers you can apply color and tonal adjustments to your images without permanently changing their pixel values. Adjustment layers are nondestructive, continuously editable, and add little to the file size of the image. Unlike "static adjustments" made via the Image > Adjustments submenu, adjustment layers stay editable. What's more, you can hide a layer's effect in areas of the image by editing its mask—plus you can restack, hide, and delete adjustment layers, as well as drag-copy them between files. You can ditch your changes and go back to the original image at any time.

Adjustment and Fill Layers

You can find adjustment layers easily:

- In the Adjustments panel, click an adjustment icon.

- At the bottom of the Layers panel, use the Adjustment Layer menu.

- From the menu, choose Layer > New Adjustment Layer (**FIGURE 9.1**).

Adjustment layers come with a layer mask, which starts out empty (white), meaning that the adjustment applies to all the layers below it. (If you have an active selection when you add an adjustment layer, the unselected area appears in black on the layer mask.) Using a Brush tool or the Gradient tool, you can paint on the mask in black where you don't want the adjustment to reach the image. The sections that follow detail working with the specific types of adjustment layers, while the sidebar "Changing Images with Adjustment Layers" provides an overview of tips that are applicable to all adjustment layers.

TIP If you want an adjustment to apply to less than 50% of the image, start with a black mask (click Invert on the Properties panel or apply Edit > Fill, Use: Black), then paint on the adjustment layer's mask in white to reveal the adjustment.

TIP If you want adjustment layers without masks, deselect Add Mask By Default in the Adjustments panel menu.

FIGURE 9.1 The Adjustments panel menu lists all the available adjustment layers, grouped thematically into three broad categories: tone, color, and (starting with Invert) radically changing pixels. The same adjustments are shown in icon form in three rows on the panel itself.

Changing Images with Adjustment Layers

Adjustment layers are nondestructive, so. you can experiment with settings and re-edit the adjustment at any time. Here are some tips for getting the most from adjustment layers:

- Reduce the effect of an adjustment layer by lowering its opacity in the Layers panel.
- Show/hide the effect of an adjustment layer by toggling the Visibility icon in the Properties panel (**FIGURE 9.2**) or Layers panel.
- Toggle the most recent edit off and on, by clicking and holding View Previous State (👁) on the bottom of the Properties panel or press and hold the \ key, then release.
- Restore default settings to an adjustment layer by clicking Reset To Adjustment Defaults (↩); the last chosen settings are restored. Click the button again to restore the default settings for that adjustment type.
- Restrict the adjustment to part of an image by painting on the adjustment layer's mask.
- Create a new adjustment with a mask of a specific shape by starting with a selection. When you add the adjustment layer the selection restricts the new adjustment with a layer mask.
- Create a new adjustment by starting with a path, then Ctrl/Command-click the Add Layer Mask button. The path restricts the new adjustment with a vector mask.
- Restrict the effect of an adjustment layer to the layer directly beneath it by clicking Clip To Layer ⬚) on the Adjustments panel. The adjustment layer becomes indented, and the underlying layer name has an underline. Or, Alt/Option-click the line between the adjustment layer and the layer below it.

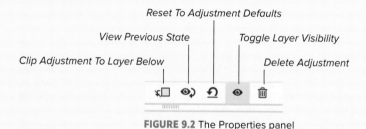

FIGURE 9.2 The Properties panel

Making Brightness/ Contrast Adjustments

In the Properties panel for a Brightness/ Contrast adjustment layer, moving the Brightness slider to the right brightens the highlights in an image; moving the slider to the left darkens the shadows. Likewise, move the Contrast slider left for less contrast, right for more contrast. Values range from −150 to +150 for Brightness, and from −50 to +100 for Contrast.

In normal mode, the adjustments are proportionate (nonlinear)—like Levels and Curves adjustments. When you turn on the Use Legacy option, all pixel values are shifted higher or lower when you adjust brightness. This can cause clipping or loss of image detail in highlight or shadow areas, so it isn't recommended for photographic images (**FIGURE 9.3**).

FIGURE 9.3
The Brightness and Contrast adjustment layer

Making Levels Adjustments

Along with Curves, Levels is one of the most useful adjustments for editing the tone and color of your images. The main advantage of Levels is that it gives you a histogram of the image. As you adjust the Levels histogram, keep an eye on how the changes affect the appearance of the histogram on the Histogram panel.

To adjust tone and color using Levels:

1. To increase the contrast of your image drag the Black and White Input Levels sliders to the edge of the first group of pixels at either end of the histogram. Move the White slider to the left to brighten the highlights; move the Black slider to the right to darken the shadows (**FIGURE 9.4**).

   Pixels to the left of the Black slider become black; pixels to the right of the White slider become white. For example, if you move the Black slider to the right at level 10, all the pixels at level 10 and lower are mapped to level 0. Similarly, if you move the White slider to the left at level 240, all pixels at level 240 and higher are mapped to level 255. The pixels in each channel are adjusted proportionately to avoid shifting the color balance.

2. (Optional) To identify areas in the image that are being clipped (completely black or completely white), hold Alt/Option as you drag the White Input Point slider to the left. Release when the first areas of color or white display; those pixels are given the lightest tonal value.

3. (Optional) To find the darkest and/or lightest pixels in the image, hold Alt/Option and drag the Black Input Point slider right or the White Input slider left. The first areas of color or black/white that appear are the darkest/lightest pixels in the image. Alternatively, choose Show Clipping For Black/White Points

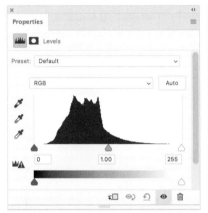

Before

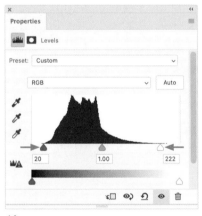

After

FIGURE 9.4 Boosting contrast with Levels, before and after moving the black and white points to the edge of the first group of pixels

from the panel menu. Restore the Black and White Point Input sliders to their starting positions, then with the Black Point eyedropper click on the image where the darkest pixels showed up. Repeat for the White Point eyedropper, clicking where the lightest pixels showed up (**FIGURE 9.5**).

4. With the shadows and highlights sliders now in their proper positions, move the middle (midtones) input slider to the left to lighten or to the right to darken the midtones.

5. Press and hold \ to toggle between the original and adjusted image, to check your adjustment.

6. (Optional) Choose a color channel from the Channel menu to adjust tones for that specific color channel.

TIP You can also enter values for the shadows, midtones, and highlights directly into the Levels fields.

TIP If the Calculate More Accurate Histogram alert button appears on the Properties panel as you're applying a Levels adjustment, click the button to refresh the histogram in the panel.

TIP To apply the current Levels settings to other images—such as to a group of photos taken under similar lighting conditions—save and apply them as a preset. There are also some ready-made Levels presets.

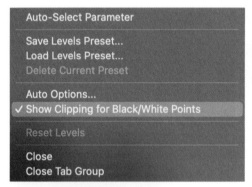

FIGURE 9.5 The results of setting the black and white points manually. The red circles indict the darkest and lightest points we chose.

To remove or neutralize a color cast:

- Select the Gray Point eyedropper then click in a part of the image that should be neutral gray (**FIGURE 9.6**). If you don't get a good result the first time, try clicking in a different area of the image.

If you want a fast "off-the-shelf" way of fixing tone and color, try using the Auto algorithms.

To auto adjust tone and color:

1. Create a Levels or Curves adjustment layer.

2. In the Properties panel, hold Alt/ Option-click Auto to access the Auto Color Correction Options.

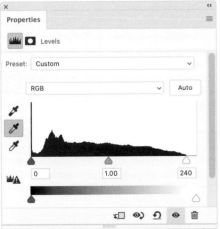

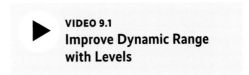

VIDEO 9.1
Improve Dynamic Range with Levels

FIGURE 9.6 Use the Gray Point eyedropper to click on an area of what should be neutral gray (the inside of the boat): see the before and after below.

3. Select your enhancement (**FIGURE 9.7**):

   ▸ **Enhance Monochromatic Contrast** clips all channels identically, making highlights lighter and shadows darker while preserving the overall color relationship. The Auto Contrast command uses this algorithm.

   ▸ **Enhance Per Channel Contrast** maximizes the tonal range for each channel, resulting in a more dramatic correction. Because each channel is adjusted individually, this algorithm may remove or introduce color casts. The Auto Tone command uses this algorithm.

   ▸ **Find Dark & Light Colors** uses the average lightest and the average darkest pixels in an image to maximize contrast while minimizing clipping. The Auto Color command uses this algorithm.

   ▸ **Enhance Brightness and Contrast** uses content-aware technology to evaluate data for the entire image, then adjusts tonal values for the composite RGB channel (instead of for each channel).

FIGURE 9.7 Before and after using the various Auto algorithms

Making Curves Adjustments

Using Curves, you can adjust specific tones, such as the highlights, quarter tones, midtones, three-quarter tones, or shadows, while leaving other areas of the image unaffected—or at least only minimally affected. Where Curves arguably has the edge over Levels is that, with practice, you can be more specific in your corrections by adding more points to the curve. The corrections can be applied to the composite image (all the channels) or to individual color channels.

An image's tone starts out represented as a straight diagonal line on a graph (**FIGURE 9.8**). With RGB images, upper-right of the line (the curve) represents the highlights. The lower-left area represents the shadows.

To adjust tone and color using Curves:

1. In the Properties panel, add a point in the top portion of the curve to adjust the highlights.

2. Move the point up to brighten, down to darken. (Notice how the straight diagonal line now curves based on the point's position.)

3. Add a point to the center of the curve to affect the midtones.

4. Add a point to the bottom portion of the curve to adjust shadows. As you add points to the line and move them, the shape of the curve continues to change, reflecting your adjustments.

TIP For a CMYK image, the controls work in the opposite way to how they do with RGB images. For CMYK images, the graph displays ink percentages. Moving the curve up adds more ink and so results in a darker images.

TIP You can also apply Curves to LAB or Grayscale images, for which the graph displays light values.

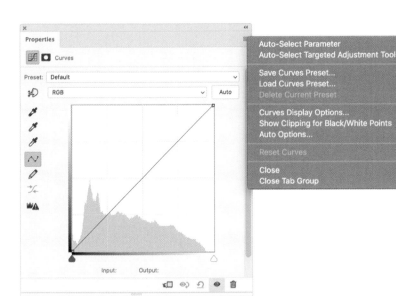

FIGURE 9.8
Making a Curves adjustment

To increase contrast using Curves:

Do one of the following:

- Use the input sliders to set the white and black tonal values. To darken the shadows and brighten the highlights, drag the Black and White Highlight Input sliders inward so they align with the ends of the gray areas in the histogram—or until you see that the contrast in the image is increased. The steeper the curve the stronger the contrast (**FIGURE 9.9**).

- To establish the white point and black point, Alt/Option-drag the Black slider until a few areas of color or black appear; do the same for the White slider until the first white pixels appear. You can also choose Show Clipping For Black/White Points from the panel menu, but most of the time you do not want this selected, so it's preferable to use Alt/Option-drag.

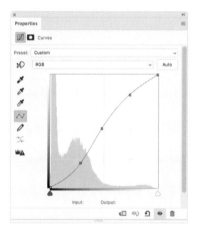

Before

After

FIGURE 9.9 Applying a contrast curve, before and after. Note we also painted on the layer mask to protect those areas from the effects of the curve.

Curve Display Options

Control what information is displayed by choosing Curves Display Options on the Properties panel menu (**FIGURE 9.10**).

Light (0–255) displays levels from 0 (black) to 255 (white). Moving the curve up adds more light and the image gets lighter.

Pigment/Ink % displays ink percentages for images in a range from 0 to 100. Moving the curve up adds more ink and the image gets darker.

Grid displays either a 4 × 4 or a 10 × 10 grid.

Channel Overlays superimposes the curve of an individual color channel's adjustments on the composite curve.

Histogram shows the image's histogram behind the curve.

Baseline shows the original 45° angle line for reference.

Intersection Line shows horizontal and vertical lines when you move control points to help you align them to the histogram or grid.

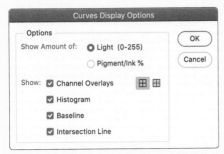

FIGURE 9.10 The Curves Display Options dialog

To make targeted adjustments with Curves:

Do one of the following:

- Add more points to adjust different tonal areas. You can affect one tonal area while leaving others relatively unchanged. If you need to remove a control point, drag it off the graph.

- Drag part of the curve; a (selected) point appears on it. For example, to lighten the midtones, drag the middle of the curve upward, or to darken the midtones, drag the middle of the curve downward. When a tonal value is lightened, its Output value is increased above its Input value (the original brightness value). When a tonal value is darkened, its Output value is reduced below its Input value.

- Click the Targeted Adjustment tool (⤲) in the Curves panel, then move the pointer over an area of the image that contains the tonal level you want to darken or lighten; a small circle appears on the curve. Click the Targeted Adjustment tool in the image to add control points to the curve. Drag upward to lighten that level or downward to darken it. A point corresponding to that level appears on the curve. Keep the Curves controls showing on the panel.

To adjust color balance with Curves

Do one of the following:

- Adjust the curves for one or more of the individual color channels.

- You can also correct the overall color by selecting the Gray Point (middle) eyedropper on the Properties panel for Curves, then click an area of the image that should be neutral gray.

TIP If the **Calculate More Accurate Histogram** alert button appears on the Properties panel as you're applying a Curves adjustment, click the button to refresh the histogram in the panel.

VIDEO 9.2
Adjust Contrast and Color Correction with Curves

Curves Shortcuts

Here are some useful shortcuts when working with Curves:

- To enlarge the graph and curve, drag an edge of the Properties panel.

- To set a point on the curve for the current channel, Ctrl/Command-click in the image.

- To change the gridline increment, Alt/Option-click the grid.

- Ctrl-Shift-click/Command-Shift-click in the image to set a point on the curve in each color channel (but not in the composite channel).

- Shift-click to select multiple points.

- Ctrl /Command+D to deselect all points on the curve.

- Press + to select the next higher point on the curve; press – to select the next lower point.

- Move selected points on the curve by pressing the arrow keys.

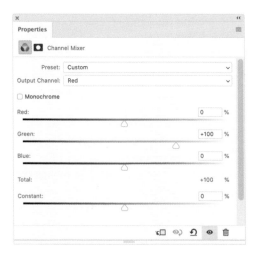

FIGURE 9.11 The result of swapping channels in the Channel Mixer. The red channel is 100% green, the green channel 100% blue, and the blue channel 100% red. A layer mask limits the adjustment to the right side of the image, showing the before and after.

Using the Channel Mixer

Use the Channel Mixer adjustment to create grayscale, sepia tone, or other tinted images (our favorite way to create grayscale images is to use a Black & White adjustment layer). Exploring the presets is a good way to start. Deselect the Monochrome option if you want to make creative color adjustments by blending one or more of its color channels (**FIGURE 9.11**).

To swap channels in the Channel Mixer:

1. In the Properties panel, choose an output channel from the Output Channel menu. This sets the source slider for that channel to 100% and all other channels to 0%. If you choose Red, for example, it starts out as 100% for Red, and 0% for Green and Blue.

2. Drag the sliders or enter a value between −200% and +200% in the field to change the amount of the color in the channel. Typically, you want the combined totals of the source channels to equal 100%. If the combined values add up to more than 100%, an icon appears to warn you that the resulting image will be brighter than the original, possibly removing highlight detail.

3. (Optional) Drag the Contrast slider to adjust the grayscale value of the output channel. Negative values add more black; positive values add more white.

Exporting Color Lookup Tables

If you want to share a look across multiple images, color lookup tables (CLUTs) can help. Specifically, you can export a CLUT from one image, and then apply it to other images that you want to have the same style (**FIGURE 9.12**). The CLUT replaces each color in an image with its corresponding color in that lookup table. Many applications support CLUTs, especially in video and 3D, so it's an easier way to get the same look across many applications used for a multimedia project.

To export and apply a color lookup table:

1. Create your desired look with a combination of adjustment layers.

2. Put the adjustment layers into a layer group, select the group, then select File > Export > Color Lookup Tables.

3. In the Export Color Lookup Tables dialog, enter a description and, optionally, copyright information.

4. Enter a value in the Grid Points field (0–256). A higher value creates higher quality, bigger files. 64 (High) is practically indistinguishable from 256 (Maximum) but is a lot quicker to save.

5. Select one or more of the available formats in which you want to export the color lookup table. We use ICC Profile, which is cross-platform and works with various color modes.

6. Specify where to save the generated files and enter a base filename to which Photoshop automatically appends the file extensions.

7. Open a new image to which you want to apply the color lookup table.

8. Add a Color Lookup adjustment layer.

9. On the Properties panel select Device Link, and load the saved lookup table.

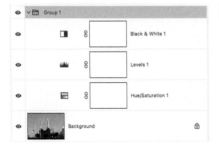

👁	✓ 🗁 Group 1		
👁	◑ 🔗 ⬜	Black & White 1	
👁	⩊ 🔗 ⬜	Levels 1	
👁	▦ 🔗 ⬜	Hue/Saturation 1	
👁	🖼 Background		🔒

Export Color Lookup Tables

Description: _23A8270.CR2　　　　　　　(OK)

Copyright:　　　　　　　　　　　　　(Cancel)

☐ Use lowercase file extensions

Quality

Grid Points:　64　High ▾

○────────────────

Formats

☐ 3DL

☐ CUBE

☐ CSP

☑ ICC Profile

Properties

▦ 🔳 Color Lookup

○ 3DLUT File　Load 3D LUT... ▾

○ Abstract　Load Abstract Profile... ▾

◉ Device Link　_23A8270.ICC ▾

☑ Dither

FIGURE 9.12 Create a "look" by combining adjustment layers. Export the adjustments layers as a CLUT and apply it to other images.

Limiting Adjustments with Clipping Masks

A *clipping mask* lets you limit the effect of an adjustment layer to just the layer directly beneath. The masking is determined by the content of the layer below. The opaque content on the base layer clips (reveals) the content of the layers above it in the clipping mask. All other content in the clipped layers is masked out (**FIGURE 9.13**).

To create a clipping mask:

1. Make sure the base layer is below the layers that you want to mask.

2. Click a layer to be clipped, or Shift-click multiple layers to be clipped (not the base layer), then press Ctrl+Alt+G/Command+Option+G.

3. To clip one layer at a time, Alt/Option-click the line between two layers (pointer), or right-click one of the layers to be clipped (not the base layer) and choose Create Clipping Mask. Alternatively, choose Layer > Create Clipping Mask.

To release a clipping mask:

-   Alt/Option-click the line separating the layers, or choose Release Clipping Mask from the Layers panel menu. Any masked layers above the clipping mask are also released.

TIP Another way to restrict the effects of an adjustment layer to specific image layers is to use a group: Select the image layers, choose Layer > New > Group From Layers. Make sure the adjustment layer is at the top of the layer group. Change the Mode from Pass Through to any other blending mode.

FIGURE 9.13 The Levels adjustment layer at the top of the layer stack is clipped to the seagull layer below, so that it affects only that layer. Without the clipping mask, the colors of the circle shape and the background would also be affected.

VIDEO 9.3
Limit Adjustments with Clipping Masks

Using Hue/Saturation Adjustment Layers

Hue/Saturation adjustment layers let you apply hue, saturation, and lightness corrections without having to create selections. If you choose Master from the menu in the Properties panel, all colors will be adjusted. To adjust a specific color range, choose that color from the menu (**FIGURE 9.14**).

To make targeted adjustments with Hue/Saturation:

Do any of the following:

- To adjust the Hue, select the Targeted Adjustment tool (🖐), then Ctrl/Command-drag horizontally over a color area in the image.

FIGURE 9.14 Adjust the hue. Note the right side of the adjustment layer's layer mask has been filled with black to show the before and after state.

- To adjust the Saturation, select the Targeted Adjustment tool and drag horizontally over a color area in the image without holding Ctrl/Command (**FIGURE 9.15**).

- With the Eyedropper click or drag in the image to select a color range. To add to the range hold Shift (or use the Add to Sample Eyedropper); to subtract from the range hold Alt/Option (the Subtract From Sample Eyedropper tool).

- For Lightness, drag the slider right to increase the lightness (add white) or left to decrease it (add black).

The color bars at the bottom of the Hue/Saturation Properties panel are like flattened color wheels. The upper bar shows the color before the adjustment; the lower bar shows how the adjustment affects all the hues at full saturation.

FIGURE 9.15
Selective Saturation with Hue/Saturation. Choose Yellows and move the Saturation slider to the right.

To make adjustments with the color bars:

Do any of the following:

- Drag the center area to move the entire adjustment slider to a different color range.

- Drag the area between the inner and outer sliders to adjust the range.

- Drag the outside sliders to adjust the amount of color fall-off (feathering of the adjustment).

TIP By default, the range of color selected when you choose a color component is 30° wide, with a 30° fall-off on either side. Setting the fall-off too low can produce banding in the image.

To colorize an image with Hue/Saturation:

1. To colorize a grayscale image you first need to convert it to a color mode. Choose Image > Mode > RGB Color.

2. Apply a Hue/Saturation adjustment.

3. In the Properties panel, select Colorize. If the foreground color is black or white, Photoshop converts the image to a red hue (0°). If the foreground color is not black or white, the image is converted to the hue of the current foreground color. The lightness value of each pixel does not change.

4. Use the Hue slider to adjust the color to your taste (**FIGURE 9.16**).

FIGURE 9.16 Colorizing an image with Hue/Saturation

Adjusting Color Balance

You can use a Color Balance adjustment to warm or cool an image or, conversely, to neutralize an unwanted cast—although Levels and Curves both offer easier ways of doing this. Choose a tonal range to adjust (Shadows, Midtones, or Highlights), then move any of the sliders toward a warmer or cooler hue. Each slider pairs a cool hue with a warm one. As you move towards one color you move away from its complement. For instance, moving towards green, reduces magenta, and vice versa. To maintain the tonal balance of the image select Preserve Luminosity (**FIGURE 9.17**).

FIGURE 9.17 A combination of Color Balance (applied to the highlights) and Vibrance to increase the saturation of the colors in the background of the image. A layer mask is applied to the layer group containing both adjustment layers to show the before and after.

Making Vibrance Adjustments

Useful for perking up flat skies and for preventing skin tones from becoming over saturated, Vibrance increases the saturation of muted colors more than the colors that are already saturated. If you want to apply the same amount of saturation to all colors regardless of their current saturation, move the Saturation slider. Sometimes this may produce less banding than the Saturation slider in the Hue/Saturation Adjustments panel (**FIGURE 9.18**).

Working with Photo Filter

The Photo Filter adjustment layer simulates the effect of using colored lens filters on your camera to cool or warm a scene or to alter the tonal response of black-and-white to a color scene. For example, if you are emulating black-and-white photography, start with a color image and add a Photo Filter adjustment layer set to Red (to deepen blue skies and increase contrast), then use that as the basis for your conversion to black-and-white.

FIGURE 9.18 Before (left) and after: Blue skies all day long with the Vibrance adjustment.

Photoshop offers a good number of preset tints, or you can use a custom color. For something subtle, pick a color similar to those in the image. Or for a more graphic effect, pick a complementary color (diagonally opposite on the color wheel) to the predominant color in the image. To further fine-tune the result, you can adjust the Density slider. Turning on the Preserve Luminosity option preserves the overall brightness and contrast values of the image (**FIGURE 9.19**)

Using Presets

Some adjustments can be captured as presets and applied to other images—useful if you have several images that require the same treatment.

To save adjustment settings, choose Save Preset from the Properties panel menu.

To apply a saved preset select it from the Preset menu of the Properties panel, or choose Load Preset from the Properties panel menu (**FIGURE 9.20**).

FIGURE 9.19 Applying Photo Filter: A layer mask is applied to the adjustment layer to show the image before (left) and after application of a cooling Photo Filter.

FIGURE 9.20 Saving and loading a predefined preset

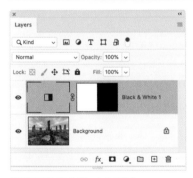

Applying Black & White Adjustment Layers

The Black & White adjustment gives you full control over how to convert a color image to a grayscale image.

To adjust tones with Black & White:

Do one of the following in the Properties panel:

- Move the sliders to the right for lighter grays; move to the left for darker grays.

- Select the Targeted Adjustment tool, then drag to the right in an area of the image to lighten that shade, or to the left to darken it. The slider corresponding to the predominant color in that area will be affected.

- Choose a preset from the menu (**FIGURE 9.21**)

To apply a color tone to a grayscale image:

1. In the Black & White adjustment layer's Properties panel, turn on the Tint option.

2. Click the color swatch, and select a color in the Color Picker.

VIDEO 9.4
Custom Black-and-White Conversions

Tinting with Gradient Map Adjustments

The Gradient Map adjustment maps the grayscale values of an image to the colors of your chosen gradient fill. With a two-color gradient fill, the shadows in the image are mapped to the starting (left) color of the gradient, and the highlights are mapped to the ending (right) color of the gradient. The midtones are mapped to the gradations in between (**FIGURE 9.22**).

To tint an image via a Gradient Map adjustment:

1. Click the gradient ramp to edit the gradient.

2. Load a preset library (or two), including the Photographic Toning library from Legacy Gradients.

3. Click a preset on the picker.

4. To minimize banding on print output, check Dither.

5. (Optional) Edit the gradient by adding or changing some color stops, or re-positioning some midpoint diamonds.

6. (Optional) Turn on the Reverse option to apply the lightest colors in the gradient to the darkest values in the image and vice versa, creating a film negative effect.

7. (Optional) Experiment with the blending mode of the Gradient Map layer. Try Multiply, Soft Light, Pin Light, Hue, or Luminosity.

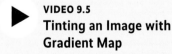
VIDEO 9.5
Tinting an Image with Gradient Map

FIGURE 9.22 Experimenting with gradient maps: magenta-orange (reversed), a more photographic dark to light gray (reversed), and the original color image

Adjusting Color with Selective Color

Selective Color adjustment layers give you another way of tweaking the colors in your image. Specifically, they're intended for a process printing scenario when you are working with CMYK images and let you modify the amount of a process color in any primary color *selectively*. For example, you can decrease the cyan in the greens while leaving the cyan in the blues unchanged (**FIGURE 9.23**).

In the Properties panel, select the color you want to adjust and drag the sliders to increase or decrease the amount of process color for that color. Selective Color uses two methods:

- **Relative** changes the amount of cyan, magenta, yellow, or black by its percentage of the total. For example, if you start with a pixel that is 50% magenta and add 10%, 5% is added to the magenta (10% of 50% = 5%) for a total of 55% magenta. (You can't use this option to adjust pure specular white because it contains no color.)

- **Absolute** adjusts the color in absolute values. For example, if you start with a pixel that is 50% cyan and add 10%, the cyan ink is set to a total of 60%.

FIGURE 9.23 Making the bluebells blue: The left side of the image shows the before state, the right side shows the adjustment achieved by increasing the cyan in the Magentas, Blues, Cyans, and White.

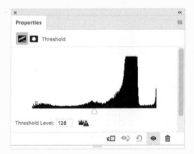

Other Adjustment Layers

These adjustments tend to get used less often—at least in our workflows.

Invert converts your image to a negative. The hue value of every pixel moves 180° around the color wheel.

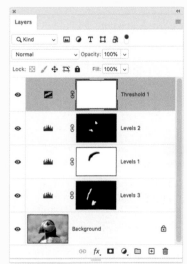

- **Threshold** creates high-contrast, black-and-white images, where all the pixels are either black or white. By default the threshold level is 128, but you can change this. All pixels lighter than the threshold number become white; all pixels darker become black (**FIGURE 9.24**).

- **Posterize** limits the number of levels and then maps pixels to the closest matching level. For example, choosing 4 tonal levels in an RGB image gives 12 colors: 4 for each of the color channels. Posterize is intended for creating special effects, such as large, flat areas in a photograph. It rarely works well by itself, but when combined with masking it can, it can be quite effective on images that have large areas of flat color (**FIGURE 9.25**).

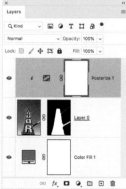

FIGURE 9.25 Posterizing the image initially resulted in ugly banding in the sky. The solution was to mask the sky, add a color fill layer of solid blue beneath, and then apply the posterizing only to the sign, which already has a limited range of colors. The adjustment is clipped to the image layer, so that it does not affect the color fill layer.

FIGURE 9.24 Used by itself, the Threshold adjustment has all the subtlety of a jackhammer. Used in combination with other adjustments (in this case Levels), it can create effective line art.

Using Fill Layers

There are three types of fill layer: solid color, gradient, and pattern. Unlike adjustment layers, fill layers don't affect the layers underneath them (unless you change their blending mode or opacity). Being non-destructive, fill layers are easier to edit and experiment with than adding static layers created with the Fill command or the Gradient tool. For example, if you want to change the angle of a gradient fill layer, you can do so instantly rather than having to recreate the gradient on a pixel layer.

Correcting Images

Here is a general workflow to follow when correcting the tone and color of an image:

1. Use the histogram to evaluate the tonal range of the image.

2. Use adjustment layers to correct the color balance to remove unwanted color casts or to correct oversaturated or undersaturated colors.

3. Adjust the tonal range with Levels or Curves adjustments. Start by setting the white and black points. This sets an overall tonal range for the image, redistributing the midtone pixels. You can then, if necessary, adjust the midtones manually.

4. As one of the last steps, apply a Smart Sharpen or Unsharp Mask filter to sharpen the edges in the image. The amount of sharpening required will vary according to the size of the image in pixels, the nature of the image, and your own preference.

5. (Optional) Target the image for printer or press characteristics.

By default, solid color fill layers apply the current foreground color to the adjustment layer. The Color Picker opens, however, so you can select a different fill color.

For gradient fill layers, you can select a gradient preset or click the gradient to open the Gradient Editor and create your own. You can also adjust the style, angle, scale, and orientation of the gradient, as required. The Dither option reduces banding by applying dithering to the gradient. Align With Layer uses the bounding box of the layer to calculate the gradient fill. You can drag in the image window to reposition the gradient.

For pattern fill layers, a pop-up panel appears where you can select a pattern, as well as enter a value for its scale and angle. Select the Snap To Origin option to make the pattern start at the same place as the document, and select Link With Layer if you want the pattern to move with the layer. With this option selected, you can drag in the image to position the pattern. Photoshop comes with various preset patterns, but it's also easy to create your own.

To create a pattern:

1. With the Rectangle Marquee tool, select an area of an image to use as a pattern. Make sure Feather is set to 0 pixels.

2. Choose Edit > Define Pattern.

3. Enter a name for the pattern.

Evaluating Images

Before working on an image, you need to evaluate it. The best way to do this—beyond the subjective evaluation of your eyes—is to use the Histogram panel. A histogram gives a quick picture of the tonal range of the image and can help determine appropriate tonal corrections. The shadows are on the left, the midtones in the middle, and highlights are on the right. The horizontal axis represents the grayscale or color levels between 0 and 255; the vertical bars represent the number of pixels at specific color or tonal levels.

While there is no "right way" for a histogram to look, an *average-key* image has detail concentrated in the midtones. An image with full tonal range has some pixels in all areas, and the overall contour of the graph is relatively solid and smooth. If an image lacks detail in a tonal range, the histogram will contain gaps and spikes, like teeth on a comb.

As you edit an image, the shape of its histogram changes to reflect changes to the tonal values. With nothing selected, the histogram represents the entire image; to see a histogram for a portion of the image, first select that portion.

You can view a histogram in three ways: Compact view, which is just the histogram; Expanded view, which is the histogram plus a menu with access to individual channels; or All Channels view, which shows a separate histogram for each channel (**FIGURE 9.26**).

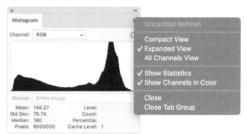

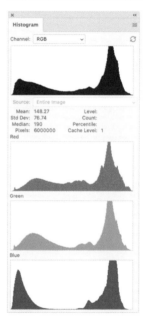

FIGURE 9.26 Examples of the three histogram views: Compact, Expanded, and All Channels (in color)

Using the Info Panel

Along with the Histogram panel, the Info panel provides useful information about your image as you make color corrections. When an adjustment layer is selected, the Info panel displays two sets of color values: *before* on the left and *after* on the right (**FIGURE 9.27**).

Choose the Color Sampler tool (or Shift-click with Eyedropper tool) to set a Color Sampler at up to four locations in the image.

Once you've added a color sampler, you can move it, delete it, hide it, or change the color sampler information displayed in the Info panel. The samplers are saved with the image, and will be there the next time you open the image.

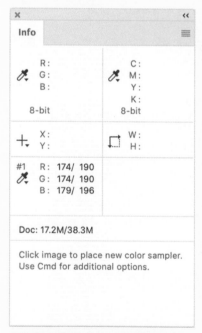

FIGURE 9.27 The Info panel showing before and after numbers for color sampler 1

To view a histogram:

Do one of the following:

- In Expanded or All Channels view, choose an option from the Channel menu: RGB (all the channels combined), a specific channel, Luminosity, or Colors.

- To display individual channels in color, choose an individual channel from the Channel menu and choose Show Channels In Color from the panel menu.

The Histogram panel reads data from the histogram cache—not from the actual image. You can set the maximum cache level (from 2 to 8) in the Performance preference. Higher cache level settings increase the redraw speed for large, multi-layer files, but require more RAM. If RAM is limited or you work mainly with smaller images, stick with lower cache level settings. Clicking the Cached Data Warning icon refreshes the histogram so that it displays all the pixels of the image in its current state.

Using Shadows/Highlights

Most of the static adjustments under the Image > Adjustments menu can be applied as adjustment layers, and it's nearly always preferable to apply this way. Some adjustments, however, are not available as adjustment layers, including Shadows/Highlights, which is particularly useful for brightening areas of shadow in an otherwise well-lit image. You can also use Shadows/Highlights to correct photos with silhouetted images due to strong back-lighting or to correct subjects that have been slightly washed from being too close to the camera flash.

The Shadows/Highlights command applies adjustments directly to the image, making it a static adjustment that discards image information. Unless that is—and here the plot thickens—you first convert your layer to a Smart Object (see Chapter 14). This allows you to apply Shadows/Highlights as a nondestructive Smart Filter.

To adjust image shadows and highlights:

1. (Optional) Convert the image layer to a Smart Object, then choose Image > Adjustments > Shadows/Highlights.

2. Move the Amount slider to adjust the amount of lighting correction or enter a value in the Shadows or Highlights percentage field (**FIGURE 9.28**).

3. If necessary, select Show More Options to access additional controls.

TIP Go easy with the Amount sliders: Extreme amounts can look unnatural.

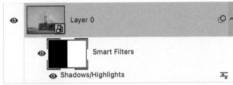

FIGURE 9.28 A Shadows/Highlights adjustment applied to the right side of the image. (A filter mask on the Shadows/Highlights Smart Filter protects the left side.)

10

Blending Modes

A layer's blending mode determines how its pixels blend with underlying pixels in the image. While some blending modes are practical workhorses used to solve specific problems, others can be used create a variety of special effects. One of the great things about using layer blending modes and opacity sliders is that there's no penalty for experimentation. Blending mode and opacity changes are nondestructive, so you can experiment without harming your data.

In this chapter, we'll demonstrate what various blending modes do, but we urge you to experiment. The possible combinations are infinite, which can be a bit overwhelming. The good news is there are no rules. If it looks good, then it is good (and vice versa).

In This Chapter

Blending Modes, Opacity, and Fill

No matter which blending mode you apply, the blending works by combining the *blend color* (applied with a painting tool or applied to a layer) with the *base color* of the layer below. What you get is the *result color*.

Along with this, the Opacity and Fill settings (found on the Options bar) determine how intensely layers blend with the layers beneath them. A layer's overall opacity determines to what degree it obscures or reveals the layer beneath it. A layer with 1% opacity appears nearly transparent, whereas one with 100% opacity appears completely opaque. The key differences between the Opacity and Fill options are:

- **Opacity** affects the whole layer, including any layer effects.

- **Fill** affects the layer transparency only, leaving the layer effect unchanged. This means that the content of the layer can be made invisible while the layer style, such as a drop shadow, can still be seen (**FIGURE 10.1**).

- **Fill** is not available when you have a group selected.

> **TIP** With the Move tool (or any tool without blending modes on the Options bar), use the number keys to quickly change the opacity of a selected layer. Press 1 for 10%, 5 for 50%, and so on. For 100% opacity, press 0.

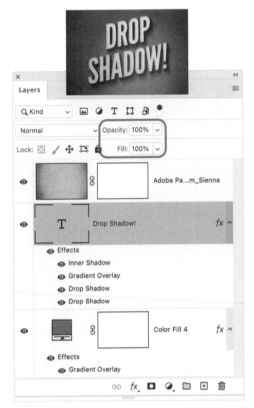 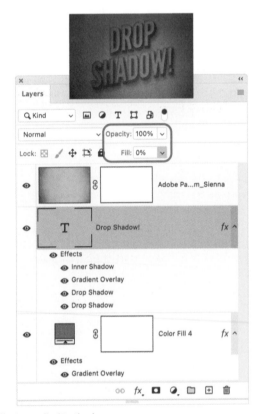

FIGURE 10.1 Fill at 100% and at 0% revealing only the effects applied to the layer.

Using Blending Modes

Combined with layer opacity, blending modes can be used to create myriad special effects. To give you an idea of what's possible, take a look at **FIGURE 10.2**.

We created a simple composition of a red rectangle as well as a black, a white, and a 50% gray square on top of an image of Battersea Power Station. Then, we applied various blending modes to the group containing these colored shapes, all at an opacity of 100%.

Normal

Multiply

Darker Color

Screen

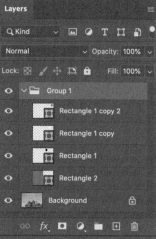

Layers panel

Lighter Color

Soft Light

Hue

Saturation

FIGURE 10.2 The same elements—a red rectangle, black square, white square, gray square—with different blending modes applied

To apply a blending mode to a layer:

1. Click any kind of layer (except the Background), multiple layers, or a group. Make sure its contents overlap some contents of the underlying layer.

2. From the menu at the top left of the Layers panel (**FIGURE 10.3**), choose a blending mode other than Normal.

3. (Optional) Adjust the layer opacity.

Because it's often hard to predict how a certain blending mode will change an image, using blending modes involves a certain amount of trial and error (**FIGURE 10.4**).

TIP With the Move tool active, press Shift++ (plus sign) to move up through the list of blending modes and Shift+− (minus) to move down through the list.

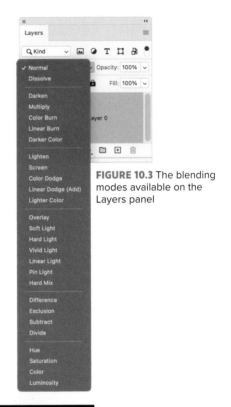

FIGURE 10.3 The blending modes available on the Layers panel

FIGURE 10.4 Overlapping different color rectangles with a range of blending modes and different opacities

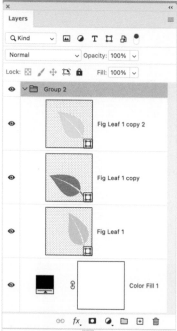

FIGURE 10.5 Changing the blending mode of the group from Pass Through to Normal, means that the blending only takes place within the group and the result is not blended with the black color fill layer, which would have resulted in a completely black image.

Default Modes

These are the blending modes you get unless you change them.

Normal

Normal is the default blending mode for a single layer: The color you paint with is the color you get. There is no blending of colors with the layers beneath, but you can get interesting results by adjusting the layer opacity.

Pass Through

Pass Through is the default blending mode for layer groups. Pass Through means that the group has no blending properties of its own—any adjustment layer, blending modes, or opacity changes applied within the group affect the way that group interacts with the layers below (**FIGURE 10.5**). Choosing a blending mode other than Pass Through means that the layers in the group are blended together first, and then that composite group is blended with the rest of the image using the selected blending mode.

Darkening Blending Modes

The blending modes in this category—Darken, Multiply, Color Burn, and Linear Burn—produce a darker result.

Multiply

Multiply is captain of the Darkening team, useful in numerous creative and workaday situations because it neutralizes white: That is, it makes the white pixels of the blend layer disappear. As well as its potential for combing two photographic images for a multiple-exposure-like effect, it is also the go-to blend mode whenever you're working with line art or handwritten text on a white background.

To create a double exposure effect using Multiply:

1. Create a layered document where the content of the top layer has a white or light subject (**FIGURE 10.6**).

2. Change the blending mode of the top layer to Multiply, allowing detail of the layer below to show through the light areas of the subject (**FIGURE 10.7**).

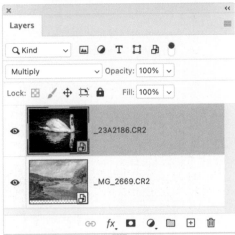

FIGURE 10.6 The original images and the Layers panel

Other Darkening Blend Modes

Darken uses the base or blend color—whichever is darker—as the result color. Pixels lighter than the blend color are replaced, and pixels darker than the blend color don't change.

Color Burn darkens and saturates. Blending with white produces no change.

Linear Burn darkens the base color by decreasing the brightness. Blending with white produces no change.

FIGURE 10.7 The resulting image. You get similar results with other darkening blending modes.

To create a composite using line art:

1. Create a layered document where the content of the top layer is on a white background—in this example a vintage line art illustration (**FIGURE 10.8**).

2. Change the blending mode of the top layer to Multiply (**FIGURE 10.9**).

To create a composite using handwritten text:

1. Select the top layer of handwritten text on a white background.

2. Change the blending mode of the layer to Multiply (**FIGURE 10.10**).

3. Because Screen is the opposite of Multiply, if you want white text, invert the layer (Ctrl/Command+I)

4. Change the layer blending mode to Screen to drop out the black (**FIGURE 10.11**).

FIGURE 10.8 Choose the Multiply blending mode to neutralize the white pixels of the layer.

FIGURE 10.10 Using Multiply drops out the white background of the poem.

FIGURE 10.9 The white of the layer "disappears" when the layer is changed to Multiply.

VIDEO 10.1
Darkening Blending Modes

FIGURE 10.11 Changing the blending mode to Screen makes the black pixels invisible

Lightening Blending Modes

Not surprisingly, the lightening group of blending modes does the opposite of the darkening blending modes, with Screen being the opposite of Multiply. While you can use Multiply to drop out or neutralize white, you can use Screen to neutralize black.

Screen

The most useful of the lightening blending modes, Screen looks at each channel's color information and multiplies the inverse of the blend and base colors. The result color is always a lighter color. Screening with black leaves the color unchanged. Screening with white produces white. The effect is similar to projecting multiple photographic slides on top of each other.

Because the black pixels of the blending layer have no effect on the layer below, using Screen is also an easy way to "select" subjects on a black background, like fireworks (**FIGURE 10.12**). If necessary, you can build density by duplicating the screened layer.

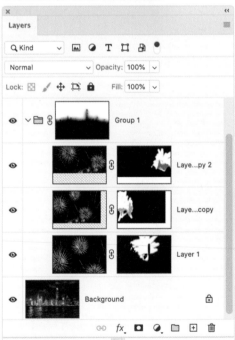

FIGURE 10.12 Missed the fireworks show? Add your own with the Screen blending mode, which will drop out the black of the fireworks layers.

FIGURE 10.13
A layer of dry ice above the Viking image.

FIGURE 10.14 Choose the Screen blending mode to neutralize the black pixels of the layer.

FIGURE 10.15
The blended result

To composite using Screen:

1. Create a layered document where the content of the top layer is on a black background—in this example a stock image of dry ice (**FIGURE 10.13**).

2. Change the blending mode of the top layer to Screen (**FIGURE 10.14**).

3. Adjust the opacity of the top layer according to taste (**FIGURE 10.15**).

 VIDEO 10.2
Lightening Blending Modes

Other Lightening Blend Modes

Lighten looks at the color information in each channel and uses the base or blend color—whichever is lighter—as the result color. Pixels darker than the blend color are replaced, and pixels lighter than the blend color don't change.

Color Dodge reduces the contrast between the blend and the base colors. The result will have saturated midtones and blown highlights. Blending with black has no effect.

Linear Dodge (Add) produces similar but stronger results than Screen or Color Dodge, brightening the base color to reflect the blend color by increasing the brightness.

Lighter Color is similar to Lighten. The difference is that Lighter Color looks at the composite of all the RGB channels, whereas Lighten looks at each RGB channel to come up with a final blend.

Contrast Blending Modes

With this category of blending modes you can increase contrast, whether you need to deepen shadows, brighten light areas, or achieve something more creative. The two most useful modes are Overlay and Soft Light.

Overlay

With Overlay, the darks get darker and the brights get brighter. Our favorite use of the Overlay blending mode is one that's not obvious: to use it as a dodge and burn layer. Dodging is selectively lightening areas; burning is selective darkening areas. While there are, as always, other ways to do this, this method is nondestructive and makes it easy to understand how and where the image has been edited.

To dodge and burn with Overlay:

1. Add a layer filled with 50% gray above your image layer.

2. Paint on this in white (with an opacity of 10–20%) to lighten the image layer (**FIGURE 10.16**).

3. Paint in black to darken the image layer. This leaves a record of the dodging and burning applied and it is completely nondestructive.

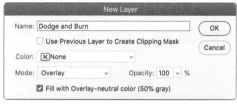

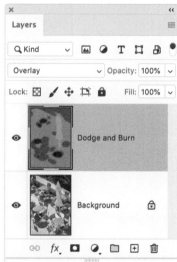

FIGURE 10.16 Applying nondestructive dodging and burning using Overlay

VIDEO 10.3
Contrast Blending Modes

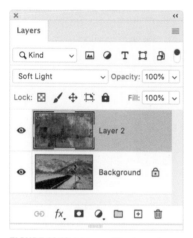

FIGURE 10.17 The source images

FIGURE 10.18 A layer of texture above an image layer.

FIGURE 10.19 The result combining the texture and image

Soft Light

Soft Light darkens or lightens the colors, depending on the blend color. The effect is similar to shining a diffused spotlight on the image. If the blend color (light source) is lighter than 50% gray, the image is lightened as if it were dodged. If the blend color is darker than 50% gray, the image is darkened as if it were burned in. Painting with pure black or white produces a distinctly darker or lighter area, but does not result in pure black or white. One of our favorite uses of Soft Light is for applying texture.

To apply texture using Soft Light:

1. Create a layered document with a texture above the image you want to affect (**FIGURE 10.17**).

2. Change the blending mode of the texture layer to Soft Light (**FIGURE 10.18**).

3. Adjust the opacity of the texture layer according to taste (**FIGURE 10.19**).

Other Contrast Blending Modes

The other contrast blending modes— **Hard Light**, **Vivid Light**, **Linear Light**, **Pin Light**, and **Hard Mix**—all dodge (lighten) and burn (darken) to varying degrees, with Hard Mix giving the strongest, most graphical results. To recite the formulas for how they work their alchemy wouldn't really give you much idea of what result to expect. For that, you just need to try them out.

Comparative Blending Modes

These blending modes are used to compare one layer to another. Of the six, we tend to reach for Difference and Exclusion more than the other four (Subtract, Divide, Hue, and Saturation).

Difference

Difference looks at the color information in each channel and subtracts either the blend color from the base color or the base color from the blend color, depending on which has the greater brightness value. Blending with white inverts the base color values; blending with black produces no change.

To manually align layers using Difference:

1. Create a document of two layers, with the layer that contains the majority of the content you want to use at the top of the layer stack (**FIGURE 10.20**).

2. Change the blending mode of the top layer to Difference (**FIGURE 10.21**).

3. Select the layer below, and with the Move tool, move the layer to align areas of detail in the layer above. For fine-tuning it may be easier to nudge with the arrow keys. In the example, we're aligning the building on the left.

4. Return the top layer to the Normal blending mode.

5. Add a layer mask to the top layer and, with black as the foreground color and a soft-edge brush, paint a hole to reveal the layer below (**FIGURE 10.22**).

FIGURE 10.20 Temporarily using Difference to align layers can help you to seamlessly combine elements of both layers. These source images were taken moments apart.

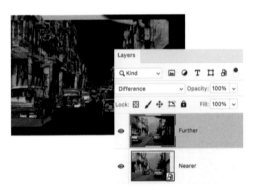

FIGURE 10.21 Choose Difference as the blending mode for the top layer. Move the bottom layer to align it with the top — aim to align the edges of the buildings.

FIGURE 10.22 Having restored the blending mode of the top layer to Normal, add a layer mask and paint on this to reveal the important part of the layer below.

FIGURE 10.23 Applying a halftone filter

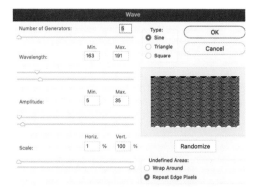

FIGURE 10.24 Adding a Wave filter

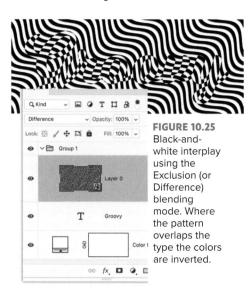

FIGURE 10.25 Black-and-white interplay using the Exclusion (or Difference) blending mode. Where the pattern overlaps the type the colors are inverted.

Exclusion

Exclusion creates an effect similar to, but lower in contrast than, Difference. Black pixels have no effect on the underlying image; white pixels invert the underlying image; and grays partially invert depending on their brightness.

Exclusion can be used to create graphic effects that use positive and negative interplay.

To create an Op Art effect using Exclusion:

1. To create a stripy pattern, first create a white color fill layer.

2. Convert the color fill layer to a Smart Object.

3. With the foreground/background colors set to black and white (press D to restore them if necessary), choose Filter > Filter Gallery > Sketch > Halftone Pattern (**FIGURE 10.23**). For Pattern Type, choose Line. Set Size and Contrast to their maximum values. Click OK.

4. Apply a Wave filter by choosing. Filter > Distort > Wave (**FIGURE 10.24**).

5. Apply a Lens Correction filter, Filter > Lens Correction, to change the angle of the pattern.

6. Create a type layer containing a single word or short phrase. Scale the type as required. Make sure it is black. Move the type layer beneath the pattern.

7. Change the blend mode of the pattern to Exclusion (or Difference) (**FIGURE 10.25**).

Divide

This blending mode looks at the color information in each channel and divides the blend color from the base color. Occasionally you can use it to create interesting graphical results that drop out black and specific colors from the image.

To create a stylized image using Divide:

1. Add a Black and White adjustment layer to an image.

2. Experiment with the presets: In this case, we chose High Contrast Red Filter (**FIGURE 10.26**).

3. Change the blending mode of the adjustment layer to Divide. This neutralizes the blacks from the image, and, because of the High Contrast Red filter, also the blue sky (**FIGURE 10.27**).

Other Comparative Blending Modes

Subtract looks at the color information in each channel and subtracts the blend color from the base color.

For details on the color-based comparisons of **Hue** and **Saturation**, see the next section.

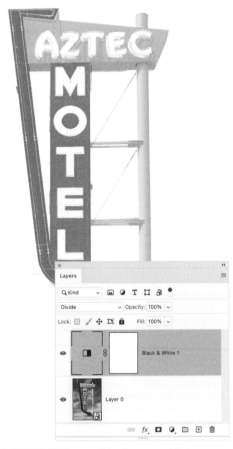

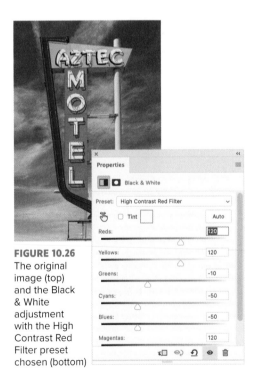

FIGURE 10.26 The original image (top) and the Black & White adjustment with the High Contrast Red Filter preset chosen (bottom)

FIGURE 10.27 The resulting image and its layers

Color Blending Modes

These blending modes affect the hue, saturation, or tonality of an image. Or, to put it another way, they affect the color, the intensity of color, or the degree of lightness and darkness of an image. Color and Luminosity are the standouts here.

Color

The Color blending mode combines the luminance of the base color and the hue and saturation of the blend color. Because it preserves the gray levels in the image, it is useful for hand-tinting monochrome images. While there are many ways to colorize grayscale images (after you've converted them to RGB) or historic photos (**FIGURE 10.28**), using layers and the Color blending mode is arguably the most intuitive.

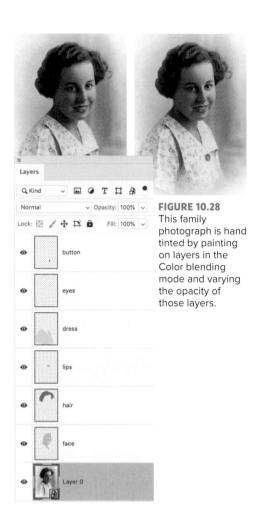

FIGURE 10.28
This family photograph is hand tinted by painting on layers in the Color blending mode and varying the opacity of those layers.

▶ **VIDEO 10.4**
Color Blending Modes

To colorize an image with Color blending mode:

1. Add layers as necessary, making their blending mode Color.

2. Choose a foreground color and paint with any of the painting tools.

3. Experiment with varying the layers' opacity as well as with the color.

To quickly change the color of the paint on a layer:

1. Lock the transparency of the layer.

2. Select a new foreground color.

3. To fill with the foreground color, press Alt+Backspace/Option+Delete. Alternatively, you can press Alt+Shift+Backspace/Option+Shift+Delete without having to first lock the layer transparency.

Luminosity

Luminosity affects the tonality of the underlying image, but not the hue or saturation. This is useful when using a Levels or Curves adjustment layer to make tonal changes to an image without introducing any change in the color. An unexpected use of Luminosity is to apply it to a Black & White adjustment layer and then use the color sliders to adjust the colors in the image (**FIGURE 10.29**).

To edit colors with Luminosity:

1. Select a color image layer.

2. Add a Black & White adjustment layer.

3. Change the blending mode of the adjustment layer to Luminosity (**FIGURE 10.30**).

4. Experiment with the color sliders to increase or decrease the intensity of a specific range of colors (**FIGURE 10.31**).

Other Color Blending Modes

Hue creates a result color with the luminance and saturation of the base color and the hue of the blend color.

Saturation creates a result color with the luminance and hue of the base color and the saturation of the blend color.

Lighter Color compares the blend and base color and displays the higher value (lighter) color.

Darker Color compares the blend and base color and shows the darker of the two.

FIGURE 10.29 Using a Black & White adjustment in Luminosity blending mode to adjust colors

FIGURE 10.30 Change the blending mode of the Black & White adjustment to luminosity

FIGURE 10.31 Adjusting the sliders for the example image to create more drama in the sky

Fill New Layers with a Neutral Color

Some filters (such as those in the Render group) can't be applied to layers with no pixels. A way around this is to fill a layer with a layer neutral color. The neutral (invisible) color will depend on the layer's blending mode—for Multiply it is white, for Screen it is black, and for Overlay it is 50% Gray (**FIGURES 10.32** and **10.33**).

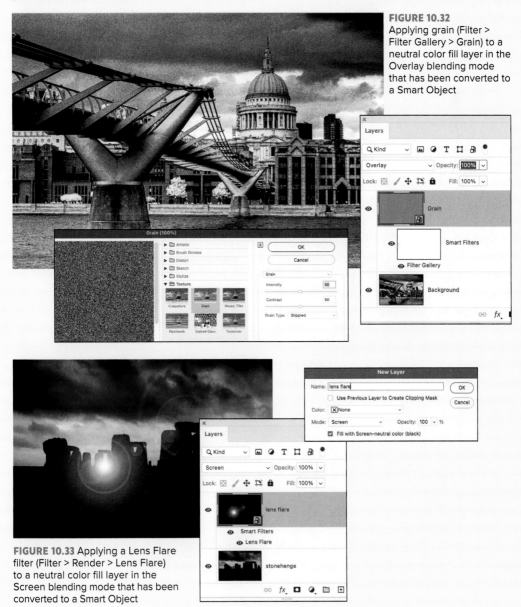

FIGURE 10.32
Applying grain (Filter > Filter Gallery > Grain) to a neutral color fill layer in the Overlay blending mode that has been converted to a Smart Object

FIGURE 10.33 Applying a Lens Flare filter (Filter > Render > Lens Flare) to a neutral color fill layer in the Screen blending mode that has been converted to a Smart Object

Blend If Options

Double-click to the right of a layer thumbnail or choose Blending Options from the Layers panel menu to bring up the Blend If options for that layer.

These sliders control which pixels from the active layer and the underlying layers are visible. For example, you can drop dark pixels out of the active layer or force bright pixels from the underlying layers to show through. You can also smooth the transition between blended and unblended areas.

A common usage of Blend If (and one that only works in certain circumstances) is to drop out the blue of the sky from the image.

To mask using the Blend If sliders:

1. Choose an image where the colors of the foreground are distinctly different from those of the sky and where the sky is a solid blue (**FIGURE 10.34**).

2. Move the the sliders at right for This Layer to the left. The further you go, the more blue will drop out (**FIGURE 10.35**).

3. To define a range of partially blended pixels, hold Alt/Option, and drag one half of a slider triangle. The two values that appear above the divided slider indicate the partial blending range (**FIGURE 10.36**).

Theoretically, using the Blend If sliders this way means you can mask a layer based exclusively on the brightness of its pixels without having to bother with selections that define your mask by area.

However, while it's fun to mess around with these options, masking "by the numbers" works only on specific types of images: those with distinct areas of pure colors. Conventional layer masking

techniques give you far more control. Furthermore, the only indication that changes have been made using the Blend If sliders is the (easy to miss) Advanced Blending badge on the layer thumbnail (**FIGURE 10.37**), which means that this approach is not as transparent (pun intended) as using layer masks, or as easy as Select Sky.

FIGURE 10.34 The original image

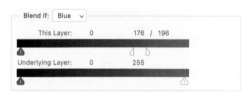

FIGURE 10.35 Using the Blend If sliders

FIGURE 10.36 The resulting image

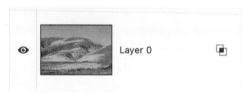

FIGURE 10.37 Layers with advanced blending options applied have a badge.

Color

When learning Photoshop, few topics are more fundamentally important than color—and the abundance of features for choosing and applying it. For example, the Swatches panel enables you to save and apply colors swiftly and efficiently, while the Color Sampler tool helps you keep track of colors in key image areas while you make adjustments.

Essential tasks involve setting the foreground and background colors, sampling color with the Eyedropper tool, as well as using the Color Picker and Color panel to mix colors in modes such as RGB, CMYK, HSB, and Lab. Duotones and spot color channels allow you to achieve results that aren't possible with CMYK printing.

To ensure your images output with the best possible (and expected) colors onscreen or in print, you need to have an understanding of a few color management tasks such as choosing appropriate color settings and working with color profiles. In this chapter, we'll cover the gamut of working with color.

In This Chapter

Choosing Colors

The foreground and background colors are Photoshop workhorses. Primarily, the foreground color is used for painting and the background color fills areas you erase on the Background layer. But, you'll also put them to work when you create gradients, format type, or apply fills and strokes to paths, selections, or shapes. Buttons for choosing foreground and background colors (■) reside in the Tools panel and the Color panel. The colors of these buttons change to reflect the current foreground and background colors. When you click these buttons, a Color Picker opens where you can choose colors using four color modes: HSB, RGB, Lab, and CMYK.

To set the foreground and background colors in the Tools panel:

Do one of the following:

- To set the foreground color, click the upper color selection box in the Tools panel and use the controls in the Color Picker.

- To set the background color, click the lower color selection box in the Tools panel and use the controls in the Color Picker.

- To exchange the foreground and background colors, click the Switch Colors icon (↰), or press X.

- To restore the default foreground and background colors (black and white), click the Default Colors icon (■), or press D.

To choose colors with the Color Picker:

Do one of the following:

- To choose colors in HSB, RGB, or Lab modes, highlight any field and enter the desired value, or drag the slider and click in the color field (**FIGURE 11.1**).

- To choose colors in CMYK mode, enter values in the CMYK fields, or move your pointer over the letters and drag the scrubby sliders that appear.

- To choose colors in hexadecimal mode, enter the code in the hexadecimal field.

- To select the nearest in gamut (▲) or web safe color (◉), click the appropriate box. The web safe color warning is a holdover from the 1990s when displays had far less color capability; today you can ignore it in almost all cases. The other warning refers to the gamut of the current CMYK working space. Be sure your Color Settings are appropriate before relying on this feature.

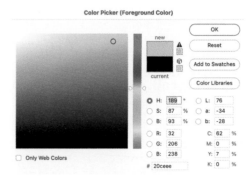

FIGURE 11.1 The Color Picker

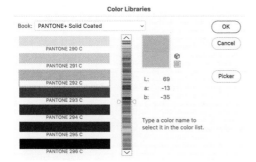

FIGURE 11.2 Choosing a spot color in the Color Libraries dialog

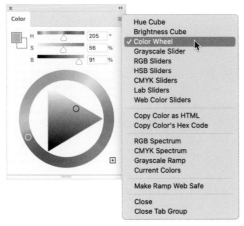

FIGURE 11.3 The Color panel offers several options for defining colors.

- To choose a color from color books such as Pantone and Trumatch, click the Color Libraries button. In the dialog that opens, choose a Book from the menu, then type the name of the specific color you want, or scroll and click a color (**FIGURE 11.2**).

- To reset to the current foreground color, click the Current color box or hold Alt/Option to change the Cancel button to Reset and click it.

To choose colors with the Color panel:

1. If the Color panel is not visible choose Window > Color.

2. In the panel menu, choose the controls you want to use: Color Wheel, Hue Cube, Brightness Cube, or sliders for the various color modes (**FIGURE 11.3**).

3. In the panel, click the foreground or background color box to select it.

4. Use the controls to choose a color.

Photoshop also offers a heads-up-display (HUD) Color Picker that you can access directly in the document window while painting.

To choose colors while painting with the HUD Color Picker:

1. Choose a painting tool.

2. Alt+Shift-right-click/Command+Option+Control-click and hold in the image to display the HUD Color Picker (**FIGURE 11.4**).

3. Release the pressed keys but keep the click held down as you drag to adjust the hue, saturation, and brightness.

4. Hold the spacebar to lock the selected hue or shade.

5. Release the click and continue painting with the new foreground color.

TIP You can change the display of the HUD Color Picker in General preferences to make it larger or smaller or use a hue wheel instead of a field.

FIGURE 11.4 The HUD Color Picker lets you choose colors on the fly while painting without moving your cursor off the canvas.

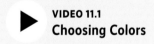

▶ **VIDEO 11.1**
Choosing Colors

Using Color Swatches

Color swatches offer a convenient way to store, organize, and apply specific colors. You can manage them in the Swatches panel.

To create, apply, and delete color swatches:

In the Swatches panel, do any of the following:

- To create a new swatch from the foreground color, click the New Swatch button.

- To use a swatch as the foreground color, click it.

- To use a swatch as the background color, hold Alt/Option and click it.

- To delete a swatch, click it, then click the Delete Swatch button.

To organize and find color swatches:

In the Swatches panel, do any of the following:

- To rename a swatch, click it and choose Rename Swatch from the panel menu.

- To group swatches, select them in the panel and click the New Swatch Group button. You can also drag swatches into and out of groups in the panel, and nest groups inside of other groups (**FIGURE 11.5**).

- To reorder swatches or groups in the panel, drag them to a different location.

- To search for a swatch, enter its name in the search field at the top of the panel.

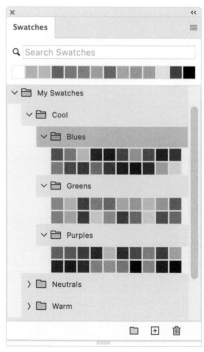

FIGURE 11.5 Keep your swatch collection organized by putting swatches into groups.

To save and export color swatches:

In the Swatches panel, do any of the following:

- To save a recently used color as a swatch, click it in the row at the top of the panel, then click the Create New Swatch button.

- To save swatches or groups for use in Photoshop, click them and choose Export Selected Swatches from the panel menu.

- To save swatches or groups for use in Adobe Illustrator or InDesign, click them and choose Export Swatches For Exchange from the panel menu. Note you can also save swatches to Creative Cloud Libraries to share them between applications.

TIP To always see the names of swatches choose **Small List** or **Large List** from the Swatches panel menu.

TIP To create a new color fill layer from a swatch, drag the swatch from the Swatches panel to the canvas.

TIP In the Color Picker, click the Add To Swatches button to immediately create a new swatch.

TIP You can add hundreds of premade swatches to the panel by choosing Legacy Swatches from the panel menu.

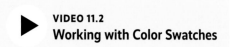

VIDEO 11.2
Working with Color Swatches

Sampling and Viewing Color Values

In Photoshop, you have two options for sampling and viewing color values from images. With the Eyedropper tool you can click to use the sampled color as your foreground or background color. With the Color Sampler tool you can place up to 10 persistent sample points on an image and view their color values using options in the Info panel. As you apply color corrections, you can see how these color values change and use that as a guide to getting the results you want.

To sample color with the Eyedropper tool:

1. In the Tools panel, click the Eyedropper tool (🖉), or press I.

2. On the Options bar, choose a Sample Size. Point Sample is the exact value of the pixel you click on. Other choices average color values from the area around where you click.

3. Choose which layers to sample colors from in the Sample menu.

4. Select Show Sampling Ring to display a large ring around the Eyedropper tool showing the sampled color above the color it is replacing as the foreground or background color (**FIGURE 11.6**).

5. To select a new foreground color, click or drag and release. To select a new background color, hold Alt/Option as you click or drag, then release.

TIP When you are using a painting tool, you can sample color without switching tools. Just hold Alt/Option to access the Eyedropper tool on the fly. Click to sample a color, and when you release you can immediately start painting with the new color.

TIP You can sample colors from anywhere on your screen, even outside of Photoshop. Click inside your image then drag without releasing until your cursor is over the color you want to sample.

TIP The Sample Size of the Eyedropper tool is also used by the Magic Wand, Background Eraser, and Magic Eraser.

To view color values with the Color Sampler tool and Info panel:

1. Display the Info panel by choosing Window > Info.

2. Select Color Sampler tool (🖉) from the Tools panel.

3. On the Options bar, choose a Sample Size. Point Sample is the exact value of the pixel you click on. Other choices average color values from the area around where you click.

FIGURE 11.6 The top half of the sampling ring shows the yellow color sampled from the sunflower.

4. Click where you want to measure color values in the image (up to 10 sample points). The value is displayed in the Info panel.

5. Drag a sample point to move it. Alt/Option-click a sample point to delete it. Click Clear All on the Options bar to remove all sample points at once.

6. (Optional) In the Info panel, click the small triangle next to a sample point to change the color mode of the values it displays (**FIGURE 11.7**). Alt/Option-click to choose a new color mode for all sample points.

By default, if you make changes to the color values in an image with adjustment layers such as Hue/Saturation, Levels, or Curves, the new color values are displayed for each sample point in the Info panel to the right of the original values (**FIGURE 11.8**). If you merge or flatten the adjustment layers, the change becomes permanent and you see only the new color sample values. You can also display just the new values without flattening or merging by choosing Panel Options from the Info panel menu and turning on Always Show Composite Color Values.

TIP You can place color sample points with the Eyedropper tool by holding Shift and clicking in an image.

TIP Sample points are saved with the image and will be retained if you close and reopen the file.

TIP The color sampler icons will disappear from the canvas if you switch to a tool other than the Eyedropper or Color Sampler, but will reappear when you choose either of those tools again.

FIGURE 11.7 Click a sample point in the Info panel to choose a different color mode for it to display.

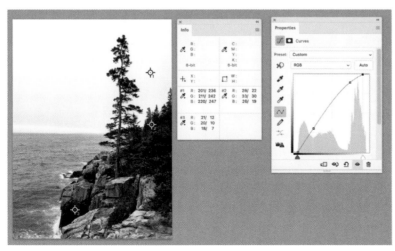

FIGURE 11.8 The three sample points on the Info panel show their original values and their values after the Curves adjustment.

Working with Duotones

Duotones are images composed of one 8-bit grayscale channel that can be printed with one to four inks. They can be used to add tints to grayscale images or extend the tonal range by printing with black and gray inks. To create a duotone, you start with a grayscale image and then map your desired ink(s) to the grayscale values with curves. Even if you don't plan to print the image, you can still make use of duotones to produce creative tinted image effects with increased dynamic range.

To create a duotone:

1. If you're starting from a color image, convert it to grayscale by choosing Image > Mode > Grayscale.

2. Choose Image > Mode > Duotone.

3. In the Duotone Options dialog, select Monotone, Duotone, Tritone, or Quadtone for the Type option.

4. For each ink, click the color box to open the Color Picker and select an ink. Click the Color Libraries button to select from color books such as Pantone. For best results, choose inks in order from darkest (Ink 1) to lightest.

5. For each ink, click the curve box and adjust the duotone curve (**FIGURE 11.9**).

6. Click OK.

7. If you plan to place the duotone in a page layout program such as Adobe InDesign, save it as a Photoshop PDF so you can preview color separations accurately in the layout later.

TIP Explore the choices in the Preset menu to see how your image would look with various inks and curves.

▶ **VIDEO 11.3**
Creating Duotones

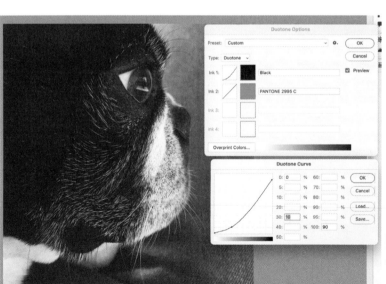

FIGURE 11.9
This duotone image of the dog is lightened by lowering the curve for the black ink.

Working with Spot Colors and Spot Channels

Spot colors are premixed inks used in color systems, such as Pantone, that offer designers the ability to print with predictable colors at a reduced cost compared to full-color printing with process (CMYK) inks. They are commonly used in duotones and when colors must be reproduced with absolute precision and consistency, such as in corporate logos. Spot colors are also used to produce colors not achievable with process inks and to designate areas for special printing effects, such as clear varnish, foil, or embossing. On press, spot colors are printed on separate printing plates from other colors. In Photoshop, they are represented by spot channels, which can be created from scratch, converted from other channels, or merged into them.

To create a new spot channel:

1. (Optional) Select the parts of the image you want to fill with the spot color.

2. In the Channels panel menu, choose New Spot Channel.

3. In the New Spot Channel dialog, click the Color box and use the Color Picker to select a spot color. If you want to specify a Pantone color, click the Color Libraries button and use the Book menu to choose the color.

4. If you're using a spot color from a color book, the name will automatically be entered. Do not change this name or the spot color may not print correctly (or at all). Otherwise, enter a name for the spot channel.

5. Enter a Solidity value. This option controls the density of the spot color; 100% is fully opaque (**FIGURE 11.10**).

6. Click OK.

TIP **To change the name, color, or solidity of an existing spot channel, double-click its thumbnail in the Channels panel.**

If you have a selection saved as an alpha channel you can use it to create a new channel to print a spot color or other effects such as a clear varnish.

FIGURE 11.10 Creating a new channel to print the text in this image in a spot color

To convert an alpha channel to a spot channel:

1. Make sure you have no active selection by choosing Select > Deselect or pressing Ctrl/Command+D.

2. In the Channels panel, double-click the alpha channel thumbnail.

3. In the Channel Options dialog, select Spot Color.

4. Click the Color box, then choose a color in the Color Picker, or click Color Libraries and choose a custom color. Click OK to close the Color Picker. If you're using the spot channel to indicate areas for printing effects such as embossing or varnish, the color you choose does not matter, because it will not be printed.

5. Rename the channel if necessary, and click OK. The grayscale values in the channel will now be used to apply the spot color (**FIGURE 11.11**).

FIGURE 11.11 Double-clicking the alpha channel thumbnail allows you to convert it to a spot channel for uses such as indicating to your printer where to apply a varnish.

To change a document with spot color channels so it prints with process inks only, you merge the spot channel(s) with the process ones. Just remember that it is almost certain that the appearance of the file will change if the spot color cannot be represented with CMYK inks. Layers will also be flattened when you merge channels. So, always create a backup copy of the file with the spot channels before merging.

To merge a spot channel into process color channels:

1. (Optional) Double-click the channel you want to merge to open the Spot Channel Options dialog and adjust the Solidity value so the image has the desired appearance. Click OK.

2. In the Channels panel, click the spot channel you want to merge.

3. In the panel menu, choose Merge Spot Channel. The image is flattened, and the spot channel is deleted and merged with the process color channels (**FIGURE 11.12**).

Keeping Color Appearance Consistent

A full exploration of color management could fill an entire book, so a deep dive into the topic is far beyond our scope here. However, you should be aware of a few color management fundamentals.

The first is that color management is the effort to compensate for the fact that cameras, scanners, displays, printers, and other devices all have different characteristics and thus will render the same color values differently. Color management seeks to make color rendering predictable and as consistent as possible across devices.

The main tool in color management is *color profiles*, which describe the color capabilities of a device (such as a camera, scanner, or display) or a set of output conditions (such as the combination of a particular press, ink, and paper). Color profiles are used by your computer's operating system and Photoshop for four tasks: opening files, showing them on the monitor, editing, and printing/exporting.

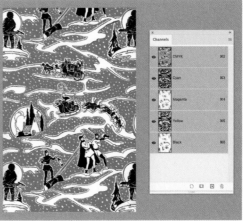

FIGURE 11.12 After merging the spot color channel, the green in this image will print with process CMYK inks.

The main point to remember is that with color profiles we can translate the color values needed to reproduce an image consistently under different conditions.

Digital cameras typically embed profiles in the images they capture. Photoshop uses these profiles to interpret the colors. But when you create a new Photoshop document for painting or compositing, or when there is no embedded profile in a document, Photoshop falls back on default profiles called *working spaces* that you choose in Color Settings. Which working spaces are best for you depend on whether your images are destined for print or screen, as well as other details of your workflow. If you have a professional print service provider, request their recommended color settings and profile conversion procedures. Otherwise, use the following steps for best results.

To choose color settings for professional print:

1. Choose Edit > Color Settings.

2. Under Working Spaces choose the following:

   ▸ RGB: Adobe RGB

   ▸ CMYK: Coated FOGRA39 or Uncoated FOGRA29, depending on whether you intend to print to coated or uncoated paper (**FIGURE 11.13**)

3. Click OK.

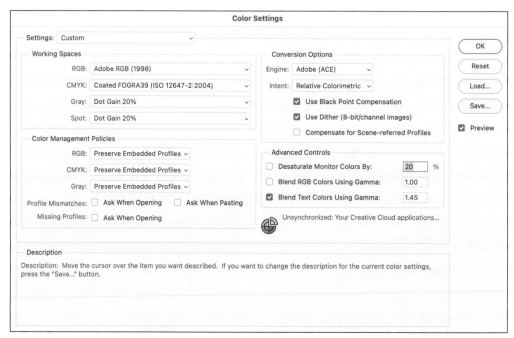

FIGURE 11.13 In lieu of Color Settings from your print service provider, Adobe RGB and FOGRA39 are good choices for print work on coated stock.

To choose color settings for web design:

1. Choose Edit > Color Settings.

2. Choose Settings: North America Web/Internet. This sets the RGB working space to sRGB, the standard for web images, and will convert RGB images to that space (**FIGURE 11.14**). When you open or paste RGB images with a different embedded color profile, Photoshop will ask you before converting those colors to sRGB. If you would rather let that conversion happen automatically, uncheck those boxes.

3. (Optional) To have more control over the conversion to sRGB and a chance to preview the effects of it on your images, change the RGB Color Management Policy to Preserve Embedded Profiles. You will need to manually convert images to sRGB at some point in your workflow, either with the Convert To Profile command or in the Save For Web dialog.

4. Click OK.

TIP Move your pointer over any settings in the Color Settings dialog to view a detailed description of what they are used for.

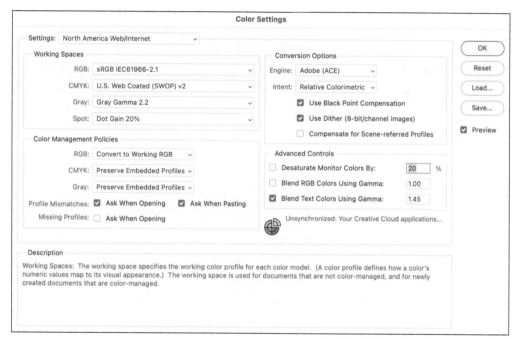

FIGURE 11.14 These Photoshop Color Settings for web work will convert images to sRGB.

Convert To Profile is the best method for converting RGB images to CMYK for print in Photoshop, when that is a requirement of your workflow. However, in many modern print workflows RGB images are preferred because they allow the print service provider to optimize the conversion to CMYK for their devices and materials. So, confirm that you actually need to deliver CMYK before converting. And in all cases, save a backup copy of the image in its original color space. Convert To Profile can also be used to convert an image from one CMYK color space to another, or from one RGB color space to another, while maintaining as much of the current appearance of the image as possible. However, colors that are out of gamut for the Destination Space will shift.

To convert to a color profile:

1. Choose Edit > Convert To Profile.

2. Under Destination Space, choose the color profile for the expected output conditions. The document will be converted to this color space and tagged with this profile (**FIGURE 11.15**).

3. Unless you have a specific reason to change any of the Conversion Options, leave them at their defaults.

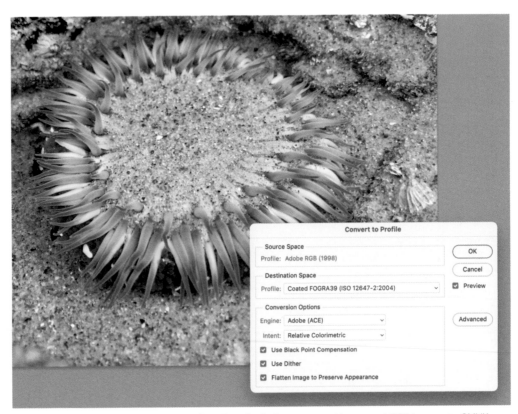

FIGURE 11.15 Convert To Profile is the preferred method when you need to convert RGB images to CMYK for print.

12

Painting

In this chapter we'll cover how to paint with solid colors, gradients, patterns, and, believe it or not, past versions of an image.

Painting with brushes in Photoshop boils down to a few basic steps. You select a tool that uses brushes, select a preset or create a new brush, then click and/or drag to paint. Brush strokes are actually composed of repeated applications of the brush tip, which can be spaced out to appear separately (e.g., a dotted line) or overlapping so closely that the effect is of one seamless stroke. The Brush Settings panel and the Options bar are home to many features for controlling how the paint is applied.

In addition to painting with the foreground color with the Brush or Pencil tool, you can use such tools as the History Brush and Art History Brush to restore parts of an image or create artistic effects from previous states in a document's history.

Other tools expand the possibilities of painting. The Gradient tool allows you to fill areas with smooth transitions of color. The Pattern Stamp tool lets you paint with patterns.

In This Chapter

Working with Brushes

There are a few fundamental tasks in working with brushes: picking a brush, choosing its options, painting (applying brush strokes), creating new brushes, and loading brushes from other sources.

You can paint with the current foreground color using either the Brush tool or the Pencil tool. The Pencil tool creates only hard-edged lines that can appear jagged and pixelated—not what you want in most cases. The Brush tool, on the other hand, allows you to apply marks with soft edges, and control the rate at which paint is applied.

To paint with the Brush tool:

1. Select the foreground color you want to paint with.

2. From the Tools panel, select the Brush tool ( ).

3. Select a brush by doing one of the following:

   ▸ Choose Window > Brushes to open the Brushes panel, and then click a brush (**FIGURE 12.1**).

   ▸ Click the Brush Preset Picker on the Options bar and click a brush (**FIGURE 12.2**).

4. (Optional) Customize the brush settings in either the Brush Settings panel or the Brush Preset Picker (see the "Brush Settings and Options" sidebar for details).

5. (Optional) On the Options bar, customize the brush options (see the "Brush Settings and Options" sidebar).

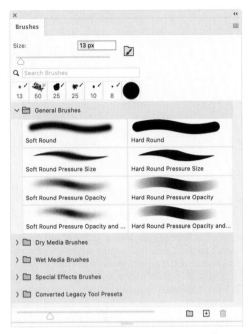

FIGURE 12.1 In the Brushes panel you can select a brush, change its size, search for brushes by name, organize them into groups, and quickly access recently used brushes and the Brush Settings panel.

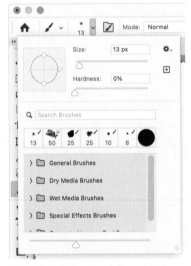

FIGURE 12.2 The Brush Preset Picker gives you the most convenient access to your brushes and related controls for modifying them.

6. Click and drag in the image. To paint a straight line at any angle, click, then hold Shift, and click again elsewhere. To paint horizontal or vertical lines, hold Shift, click, then move your cursor where you want the line to end and click again.

In addition to applying brush strokes manually with a mouse, trackpad, or pen tablet, you can also apply strokes to paths using brushes.

To paint with a brush along a path:

1. In the Layers panel, click the layer you want to paint on.

2. Set the foreground color as the color you want to paint along the path.

3. In the Brushes panel (Window > Brushes), select a brush and size.

4. Draw a path using any Pen tool or Shape tool in Path mode.

5. In the Paths panel, click the Stroke Path button (○) or press Enter/Return (**FIGURE 12.3**).

TIP To quickly reset all painting tool options to their defaults, right-click the Tool Preset Picker button (✎) on the Options bar and choose Reset Tool.

TIP When using any painting tool, you can right-click anywhere in your image area to quickly access the Brushes Preset Picker.

TIP As you're painting, you can quickly sample a color from an image to paint with by Alt/Option-clicking anywhere on the image. This reveals the HUD (Heads Up Display) Color Picker. The top half of the inner circle shows the color you're sampling. The bottom half shows the current color you're painting with.

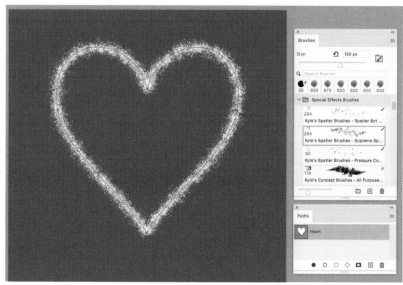

FIGURE 12.3 You can add brush strokes quickly and precisely to any shape.

Brush Settings and Options

The magic of brushes lies the many ways you can customize them. The ability to combine and fine-tune these settings and options gives you, in effect, an almost infinite variety of brushes at your fingertips. Here's a quick tour of what you can set in the Brushes Settings panel:

- **Size:** The diameter of the brush in pixels. You can change the Size value by pressing the Left and Right Bracket keys ([and]).

- **Hardness:** The percentage of the brush diameter that paints at the full Opacity value. Low percentages make soft brush strokes that taper off gradually to transparency. High values make brush strokes with clear, crisp edges (**FIGURE 12.4**). Hardness is not used in brushes sampled from images. To change the Hardness value in 25% increments, press Shift+[or Shift+].

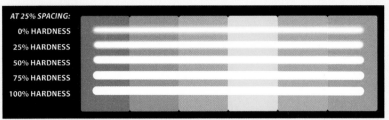

FIGURE 12.4 Various Hardness values with a 75-px round brush and default spacing

- **Roundness and Angle:** The ratio of the brush's height to its width and the angle of the brush tip relative to the canvas, respectively. Angle has no effect on strokes made with simple round brushes. In addition to entering numerical values, you can change these values interactively by clicking and dragging on the proxy. Drag the small white circles to change the Roundness. Click or drag anywhere else on the proxy to change the Angle.

- **Spacing:** The distance between the individual brush marks in a brush stroke, measured as a percentage of the brush diameter. The default value of 25% can produce lumpy, uneven brush strokes, especially when using high Hardness. Set it to 10% for smoother brush strokes at any Hardness. If you do want to see separate brush marks, increase the Spacing value. When Spacing is turned off, the spacing is determined by your cursor speed (faster moves increase spacing) (**FIGURE 12.5**).

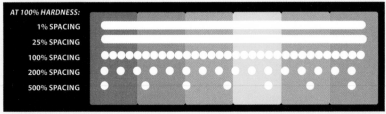

FIGURE 12.5 Various Spacing values with a 75-px round brush and 100% Hardness

- **Advanced Options:** Listed on the left side of the Brush Settings panel, includes Shape Dynamics, Color Dynamics, Scattering, Texture, and more. Click the lock icon to retain these options when you select a different brush. Select Clear Brush Controls from the panel menu to reset all.

On the Options panel, you can also select:

- **Mode:** The paint mode. Choose Normal to replace pixels as you paint over them. Choose Behind to paint on transparent areas only (so the effect is that you're painting behind existing content). Choose Clear to erase pixels to transparency as you paint. (To access Clear mode on the fly, hold tilde [~] as you paint.) Other blending modes mix the foreground color with existing content you paint over. (For more about blending modes, see Chapter 10.)

- **Opacity:** The opacity of your brush stroke. Use the slider or press single number keys to set the level of opacity in 10% increments (e.g., press 5 to paint with 50% opacity or 0 for 100% opacity). Press two number keys quickly to set a more specific percentage (e.g., press 75 to paint with 75% opacity).

- **Flow:** The rate at which the paint is applied by a brush (not available for the Pencil tool). The higher the percentage the fewer strokes needed to reach the full Opacity value. Use the slider or hold Shift and press a number key to set the Flow rate.

- **Airbrush (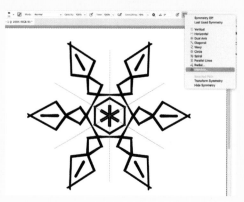):** Applies paint continuously when you click (even if the cursor isn't moving), until the full Opacity value is reached in the entire brush tip area (not available for the Pencil tool). It is most useful in conjunction with a low Flow value when you want to slowly build up the effect of your brush strokes.

- **Smoothing:** Reduces the amount of jitter in your brush strokes to give them a smoother appearance. When you increase the Smoothing value you will notice the brush stroke lags behind your cursor. With a Smoothing value of 0 the brush stroke will follow your cursor exactly. Use the slider or hold Alt/Option and press a number key to set the Flow rate. Click the gear icon to access various Smoothing modes.

- **Angle:** The angle of the brush tip (doesn't affect round brushes). Press the Left Arrow key to rotate the brush angle 1° in a positive direction and the Right Arrow to rotate 1° in a negative direction. Add the Shift key to rotate in 10° increments.

- **Auto Erase:** Paints with the background color when your brush stroke starts in an area containing the foreground color (only for the Pencil tool).

- **Symmetry:** Paints with multiple reflected copies of each brush stroke in a variety of styles (**FIGURE 12.6**). To paint with Symmetry, click the butterfly icon (🦋) and choose the type of symmetry you want. Transformation handles appear on the Symmetry Path so you can scale, rotate, or move the areas of symmetry, which are called segments. When you're done, click the Commit Transform button. The Radial and Mandala allow you to set the number of segments. You can also use any selected path that you draw with the Pen tools or Shape tools to define symmetry segments. To turn Symmetry off, click the butterfly icon and choose Symmetry Off.

FIGURE 12.6 You can make intricate designs with just a few brush strokes when painting with symmetry.

To create a new basic brush preset:

1. From the Tools panel, select the Brush tool (![brush](brush icon)). Then choose Window > Brush Settings, or click the Brust Settings button (![settings](settings icon)) on the Options bar.

2. In the Brush Settings panel, click a brush tip shape.

3. Use the controls in the panel to set basic options such as Size, Angle, Roundness, Hardness, and Spacing.

4. (Optional) Click a category on the left side of the panel to add advanced options such as Shape Dynamics, Scattering, and Texture.

5. At the bottom of the panel, click the Create New Brush button. Give your brush a descriptive name, and decide if you want to include the size, color, and tool options in the preset. Click OK (**FIGURE 12.7**).

6. (Optional) Add your new brush preset to a brush group by dragging it into an existing group. Or, click the New Group button to make a new brush group, and drag your new brush into it. Note that a brush must be in a group in order to export or import it.

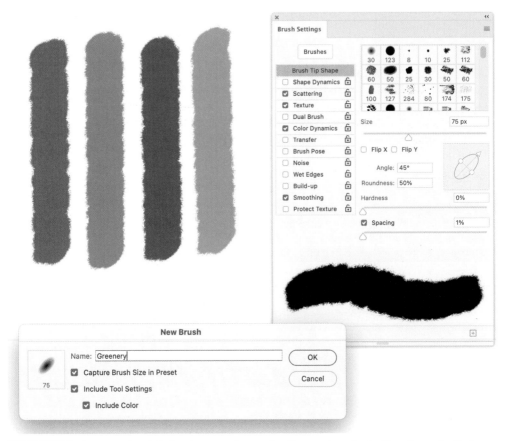

FIGURE 12.7 Saving a new brush that makes use of several options such as Angle, Roundness, Scattering, Texture, and Color Dynamics. Note the ability to also retain the size, tool settings, and color with a brush.

You can also create brush presets from images. Note that the brush does not retain any color from the sampled image; it will be a grayscale version of it and use the foreground color when you paint (**FIGURE 12.8**).

If you select a color image, the brush tip image is converted to grayscale. Any layer mask applied to the image doesn't affect the definition of the brush tip.

To create a brush preset from an image:

1. Select the image area you want to use as a custom brush tip.

2. (Optional) Press Ctrl/Command+T to transform (scale, flip, rotate) the selected content to match the way you want the brush tip to look.

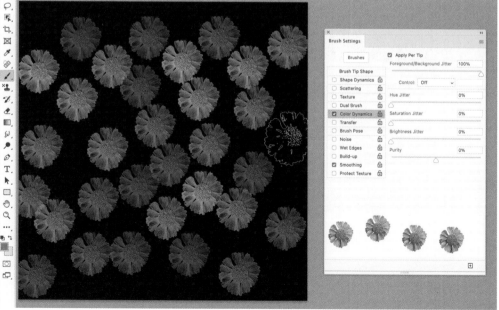

FIGURE 12.8 By creating a brush tip made from this image you can paint with flowers. Note the different colors in each brush mark come from the Color Dynamics setting, with Foreground/Background Jitter set to 100%.

3. (Optional) Feather the selection to soften the brush. The Hardness option is disabled for image brushes, so you need to feather the selection to soften the brush. Choose Select > Modify > Feather and in the dialog, enter the amount of feathering you want as the Feather Radius and click OK.

4. Choose Edit > Define Brush Preset, name your brush, and click OK.

In addition to creating your own brushes, you can also load brushes from colleagues or brushes you purchase and download from the internet. Adobe offers a vast collection of brush packs designed by the artist Kyle T. Webster as a bonus for Creative Cloud subscribers.

To load brushes and brush packs:

From the Brushes panel menu, choose one of the following:

- Import Brushes. Then navigate to the Adobe brush file (ABR) and open it. Alternatively, you can simply double-click the brush file to import it.

- Get More Brushes. This takes you to the page on Adobe's website where you can browse and download Kyle's Brush Packs. Once downloaded, double-click on a brush file (ABR) to import it.

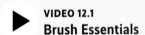
VIDEO 12.1
Brush Essentials

Using the History Brush Tool

The History Brush tool allows you to restore parts of an image to an earlier history state by painting over them.

To paint with the History Brush:

1. In the Tools panel, click the History Brush tool ( ).

2. On the Options bar, select a brush and brush options, blending mode, Opacity, Flow, and Angle.

3. Open the History panel by choosing Window > History.

4. In the History panel, click the box to the left of the state or snapshot that you want to use as the source for painting with the History Brush tool. A brush icon appears next to that state or snapshot.

5. Drag over the parts of the image you want to replace with pixels from the selected history state (**FIGURE 12.9**).

TIP Because Photoshop saves a limited number of history states, it's best to save a snapshot of a state that you want to paint from. Otherwise, you risk it disappearing from the panel as you perform more steps.

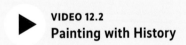
VIDEO 12.2
Painting with History

FIGURE 12.9 Running the Dust and Scratches filter instantly removes many of the defects from this antique photo, at the cost of removing important detail from the men's faces. But those details can be painted back in with the History Brush using the original state of the document (in the History panel) as the source.

Using the Art History Brush Tool

Like the History Brush tool, the Art History Brush tool lets you paint with pixels from a history state or snapshot selected in the History panel. What makes this tool unique is the way it creates stylized brush strokes by combining the data from the History panel with options you set on the Options bar.

To paint with the Art History Brush:

1. In the Tools panel, click and hold on the History Brush tool (🖉) to reveal and select the Art History Brush tool (🖌).

2. On the Options bar, select a brush and options such as blending mode, Opacity, Style (shape of the paint stroke), Area (larger values result in more applied strokes and a greater area covered with each stroke), and Tolerance (the areas where brush strokes can be applied).

3. Open the History panel by choosing Window > History.

4. In the History panel, click the box to the left of the state or snapshot that you want to use as the source for painting with the History Brush tool. A brush icon appears next to that state or snapshot.

5. Drag in the image to paint (**FIGURE 12.10**).

TIP Use a relatively small brush size (under 50 px) when using the Art History Brush tool for best results.

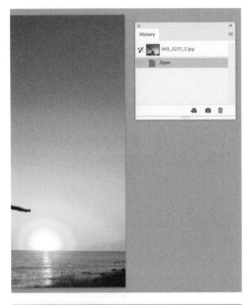

FIGURE 12.10 Use the Art History Brush to create painterly effects using a history snapshot as the source.

Using the Pattern Stamp Tool

With the Pattern Stamp tool, you can paint with a pattern instead of a solid color.

To paint with a pattern:

1. In the Tools panel, click and hold on the Clone Stamp tool (🖊️) to reveal and select the Pattern Stamp tool (🖊️).

2. Choose a brush from the Brush Presets menu on the Options bar or the Brushes panel.

3. On the Options bar, select a Pattern, then set the Mode, Opacity, Flow, and Angle to paint with.

4. (Optional) On the Options bar, turn on Aligned to paint with one seamless pattern using multiple brush strokes. Turn it off to have the pattern start at a different point with each brush stroke. Turn on Impressionist to paint with solid colors from the pattern, but no detail from it.

5. Drag in the image (**FIGURE 12.11**).

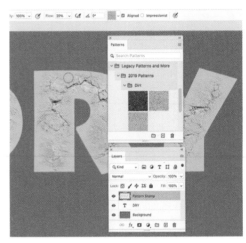

FIGURE 12.11 You can add texture exactly where you want it by painting with a pattern.

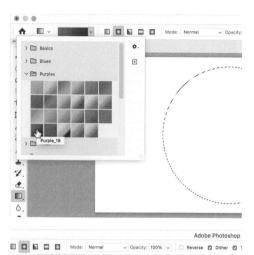

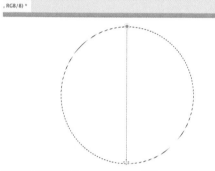

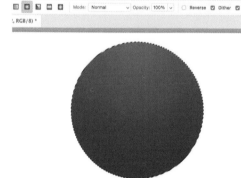

FIGURE 12.12 You can create a 3D sphere effect by dragging off center within a circular selection, using a radial gradient preset.

Using the Gradient Tool

Use the Gradient tool to fill areas in your image with blends between multiple colors, using preset gradients or ones that you make.

To apply gradients with the Gradient tool:

1. If you don't want to fill the entire current layer with the gradient, select the area you want to fill.

2. In the Tools panel, click and hold on the Paint Bucket tool (⟍) to reveal and select the Gradient tool (▤).

3. On the Options bar, select a gradient by doing one of the following:

   ▸ Click the triangle next to the gradient preview to select a gradient preset.

   ▸ Click inside the gradient preview to open the Gradient Editor. In the Gradient Editor, choose a preset or create a new gradient using the controls.

4. On the Options bar, set the remaining gradient options, such as Type, Mode, Opacity, and so on.

5. Position your pointer in the image where you want to set the starting point of the gradient and drag. Release to set the ending point. Hold Shift as you drag to constrain the angle of the gradient to a multiple of 45° (**FIGURE 12.12**).

 VIDEO 12.3
Working with Gradients

Editing Gradients

Gradients are composed of Opacity Stops, plus Color Stops and the Midpoints between them. Color Stops determine the colors that are blended in the gradient, and Opacity Stops determine the opacity or transparency at any point along the gradient.

Use the Gradient Editor to create a gradient preset by changing the color and position of stops, and the position of Midpoints. You can start from scratch or select an existing preset to begin from.

To open the Gradient Editor, click the gradient preview on the Options bar when you have the Gradient tool, or open the Gradient panel (Window > Gradients) and click the New Gradient button.

In the Gradient Editor, Opacity Stops are represented by squares and triangles attached to the top of the gradient. Color Stops look the same but are attached to the bottom. Use the following techniques to edit a gradient:

- Double-click a Color Stop to change its color.
- Drag a Color Stop or Midpoint to a different location (or set the Location numerically in the controls below).
- To remove a Color Stop, drag it off the gradient.
- To duplicate a Color Stop, Alt/Option-click it then drag the new stop to a different location.
- To change the percentage of an Opacity Stop, click it to change the Opacity value in the controls below.
- To add a Color Stop or Opacity Stop, click just below or above the gradient.
- Click New to create a new gradient preset.

The same controls can be used to modify the gradient fill of a shape layer from the Options bar or Properties panel (**FIGURE 12.13**).

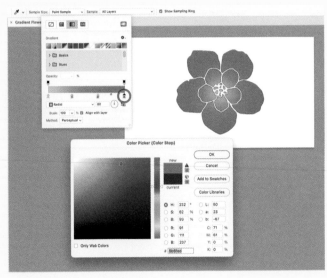

FIGURE 12.13 Double-click a color stop to open the Color Picker and select a new color for the stop.

13

Retouching

Photoshop offers several powerful methods for retouching photos. Whether your aim is to touch up a few small spots, rearrange major image details in a seamless manner, or anything in between, there's a tool for the job.

You can touch up an area with pixels from elsewhere in the image or from a separate image altogether. You can replace or repair color and detail together or separately. When you want to move an object or fill in a gap, you can take advantage of Content-Aware tools that will intelligently add in realistic details. You may need to use more than one retouching tool to fix all the defects in an image.

Regardless of which (or how many) tools you use, always work nondestructively on a new layer or duplicate layer. That will give you the flexibility to blend, modify, or remove retouched pixels at any time, and preserve the original, unretouched image.

In This Chapter

Using the Spot Healing Brush Tool

The Spot Healing Brush offers a quick way to fix small, simple defects in an image. You just pick a brush and click on (or drag over) what you want to get rid of. The tool automatically attempts to match the look of surrounding pixels for a seamless repair. It's great for removing blemishes, scratches, and other small defects in areas of relatively uniform appearance.

To fix small image defects with the Spot Healing Brush tool:

1. In the Tools panel, click the Spot Healing Brush tool ( ) to select it.

2. On the Options bar, choose a brush.

3. For Mode, choose a blending mode. Normal works best in most cases; it preserves the details of the surrounding pixels. When you only want to lighten or darken the repaired area, choose Lighten or Darken.

4. Choose a Type. Content-Aware works best in most cases as it takes into account details and edges from the surrounding area. Create Texture attempts to generate a texture inside your brush stroke. Proximity Match factors in pixels at the edge of your brush stroke only.

5. Move your pointer above the area you want to repair. Make your brush size just larger than the defect by pressing [or], then click or drag once in a short stroke. Repeat as needed to fix other defects in the image (**FIGURE 13.1**).

> **TIP** Lower the working layer's opacity for a more subtle change.

> **TIP** Show or hide the layer to assess your retouched areas compared to the original layer. Use the Eraser tool and try again if desired.

FIGURE 13.1 After setting the options, click or drag with the Spot Healing Brush to eliminate unwanted details in an image. It makes quick work of tasks like clearing the bugs off this lily.

Using the Healing Brush Tool

With the Healing Brush tool you can fix image defects and remove unwanted elements. Unlike the Spot Healing Brush tool, the Healing Brush tool lets you pick the area it uses as a source for healing (from either the current document or another open document in the same color mode). This tool also attempts to create a seamless repair by comparing the texture, lighting, and transparency in the image, and its results are usually quite good.

General Retouching Tips

Regardless of which tool you use, there are a few common best practices when it comes to retouching.

The first is to work on a duplicate layer or a new, empty layer. That way, you always have the original layer to go back to and you can reduce the opacity of the retouched copy layer to blend it with the original for a more subtle change. To make a duplicate in the Layers panel, click the layer you want to retouch and press Ctrl/Command+J.

Another best practice is to confirm that Sample All Layers is selected on the Options bar. If you're not getting the expected result (or any result at all) with one of the retouching tools, chances are turning this option on will resolve the problem.

To fix image defects with the Healing Brush tool:

1. In the Tools panel, click and hold the Spot Healing Brush tool ( ) to reveal the Healing Brush tool ( ) and select it.

2. On the Options bar, choose a brush and a blending mode. Normal works well in many cases. Use Replace to preserve more grain and texture while using a soft-edged brush. Use Lighten or Darken when the entire defect is darker or lighter than the surrounding area. Use Color when you want to change only the color (not details) of the healed area and Luminosity for the opposite.

3. Use Source on the Options bar to specify the source of the pixels used to make the repair. Sampled uses pixels from the image. Pattern uses pixels from a pattern you select from the Pattern popup panel.

4. (Optional) Turn on the Aligned option if you want the area of sampled pixels to move with your pointer with each click or brush stroke, maintaining its relative position. When it's off, pixels from the initial sample point are used at the start of each brush stroke.

5. For Sample, choose Current Layer, Current & Below, or All Layers to determine which layers are taken into account when sampling pixels. (Pixels from hidden layers are never used.)

6. Adjust the Diffusion value to determine how quickly healed pixels blend with the surrounding area. Lower values preserve fine details; higher values give a smoother effect.

7. Alt/Option-click the image area you want to sample pixels from to set the sampling point.

8. Drag with the Healing Brush in the image. When you release the drag, Photoshop uses the sampled pixels to heal the defect (**FIGURE 13.2**).

FIGURE 13.2 The Healing Brush is great for fixing cracks, tears, and other linear defects.

Using the Patch Tool

The Patch tool makes it easy to remove defects and other unwanted items from an image: Select the defect, then drag that selection over the area of the image you want to use as the replacement. You can also reverse the process and make a selection from a good area (known as the Destination) and drag it over the problem area (the Source). Use this tool to patch over stains, scratches, and spots, as well as larger items such as distracting people or objects. Using the Content-Aware option can produce results that blend in most seamlessly.

To remove unwanted details with the Patch tool:

1. In the Layers panel, click the layer with the defect.

2. (Optional) Make a selection using any of the Selection tools.

3. In the Tools panel, click and hold the Spot Healing Brush tool (🖌) to reveal the Patch tool (⬚) and select it.

4. Set the Patch mode: Content-Aware usually yields the best results and is the only way to use information from all visible layers to generate the patch (select Select Sample All Layers on the Options bar). Higher values for the Structure and Color options retain more detail and color of the Source. Normal mode yields a less seamless patch but offers unique options like filling the patch with a pattern. Choose Destination to use the initial selection area to define the patch before moving it to cover the defect.

5. Drag around the defect you want to replace (or keep your existing selection if you had one). Or, hold Alt/Option to make the Patch tool behave like the Polygonal Lasso tool, and click to make a polygonal selection with straight sides.

6. Drag the selection over an area in the image that you want to use as the patch (**FIGURE 13.3**). Or, if you chose

the Normal mode with the Destination option, drag the patch selection over the area you want to replace (**FIGURE 13.4**).

 VIDEO 13.1
Using the Patch Tool

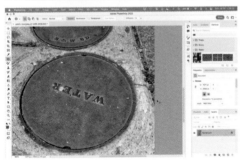

FIGURE 13.3 The Patch tools lets you remove unwanted elements by selecting and dragging them to an area containing the pixels you want to use as a patch.

FIGURE 13.4 If you would rather select a good area and drag it over the problem area, use the Destination option.

Using the Red Eye Tool

With the Red Eye tool, you can quickly fix the unsightly red pupils that occur in photos of people or animals when light from a camera flash reflects off the back of the retina. Getting good results sometimes takes a bit of trial and error but because there are only two settings, you can quickly experiment and find the right values to get the job done.

To fix red eye in people or pets:

1. Duplicate the layer containing the subject with red eyes.

2. Zoom in on the subject's red eyes.

3. In the Tools panel, click and hold the Spot Healing Brush tool (✐) to reveal the Red Eye tool (+⦿) and select it.

4. On the Options bar, choose a Pupil Size and Darken Amount. The default settings are a good place to start. For best results, you may need to use different values for each eye.

5. Click in the center of the subject's pupils. If you're not satisfied with the results, undo, adjust the values, and try again (**FIGURE 13.5**).

TIP If there is still red coloring in the subject's irises, use the Color Replacement tool to fix it.

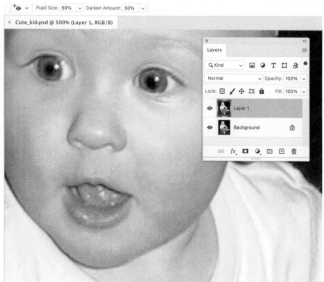

FIGURE 13.5 Eliminating red eye usually takes just a few clicks, though there can be some trial and error to dial in the best settings.

Using the Clone Stamp Tool

With the Clone Stamp tool you can duplicate elements in an image or remove them by painting, or "cloning," copies of pixels from one area to another. You can also clone pixels from one document to another, as long as those documents both use the same color mode. Unlike the Healing and Patch tools, Photoshop does not perform any automatic blending with the Clone Stamp tool.

To retouch an image with the Clone Stamp tool:

1. In the Tools panel, click the Clone Stamp tool (🔨) to select it.

2. On the Options bar, choose a Brush (soft round often works well), Mode (Normal for most jobs), Opacity (for cloned pixels), Angle (angle of the brush tip to the canvas; no effect on round brushes), and whether the tool samples from the Current Layer, Current & Below, or All Layers.

3. (Optional) Click the Clone Source Panel (🖼) icon on the Options bar to open the panel, specify sources from up to five documents, and transform pixels as you apply them.

4. (Optional) Adjust the Flow rate (rate at which cloned pixels are applied). Higher Flow means fewer clicks or drags are needed to reach the full Opacity value.

5. (Optional) Turn on the Aligned option to move the area of sampled pixels with your pointer with each click or brush stroke, maintaining its relative position. When it's off, Photoshop uses pixels from the initial sample point for each stroke.

6. Alt/Option-click on the pixels you want to clone to sample them.

7. Create a new, empty layer and keep it selected. By putting the cloned pixels on this layer you can work nondestructively.

8. Press [or] to change the brush size. A preview of the cloned source pixels appears inside the brush cursor.

9. Click or drag to apply short brush strokes of cloned pixels (**FIGURE 13.6**).

10. (Optional) To set a new source point in between brush strokes, Alt/Option-click in a different area in the source document. Changing the brush size between strokes can also help you avoid obvious evidence of cloning.

TIP To prevent obvious repeating elements (aka cloning tracks), use short strokes with the Clone Stamp tool and use multiple samples from different areas.

TIP With the Fade command you can change the opacity and blending mode of cloned pixels to better meld them with surrounding content. Choose Edit > Fade Clone Stamp immediately after cloning. It works even in a flattened document containing just a Background layer.

▶ VIDEO 13.2
Using the Clone Stamp Tool

FIGURE 13.6 Using the Clone Stamp tool requires more time and effort than other retouching tools, but in exchange it gives you more direct control over the results than other retouching methods.

Using Content-Aware Fill

Content-Aware Fill gives you a powerful yet relatively easy method for removing unwanted objects from your image. When you invoke Content-Aware Fill, Photoshop switches to a new workspace, showing only the tools and options for using this feature. You can output the results to the current layer or a new one.

FIGURE 13.7 The first step to getting rid of the bird in this image is selecting it.

To remove objects from an image with Content-Aware Fill:

1. Make a selection of what you want to remove. You can use any selection method, as long as it fully selects the unwanted object (**FIGURE 13.7**).

2. Choose Edit > Content-Aware Fill. The interface switches to the Content-Aware Fill workspace. On the left, it shows the original image with a sampling area overlay indicating where Photoshop will sample detail from in order to fill in the selected area. In the center is a preview area. On the right are controls you can use to fine-tune the appearance of the replacement pixels (**FIGURE 13.8**).

3. (Optional) Use the Sampling Area Options to refine the sampling area (see Video 13.3 for details).

FIGURE 13.8
The Content-Aware Fill workspace. Note the option to output the replacement pixels to a new layer.

4. In Output Settings you can choose to output to the current layer (not usually recommended because this is a destructive change), a duplicate of the current layer, or a new layer containing only the modified pixels.

5. Click OK to apply the Content-Aware Fill and return to your previous workspace (**FIGURE 13.9**).

VIDEO 13.3
Using Content-Aware Fill

Using the Content-Aware Move Tool

With the Content-Aware Move tool, you can reposition elements of an image in a seamless manner. When you move a selection, Photoshop automatically recomposes the image to fill in the hole with matching details from the surrounding area. In Move mode you can move people and objects to new locations, while Extend mode enables you to expand or contract objects with regular repeating elements. The results can be almost magical.

FIGURE 13.9 The result of using Content Aware-Fill: a new layer where the initial selection is filled with pixels that seamlessly hide the bird.

FIGURE 13.10 Use Content-Aware Move to reposition elements and blend them with their new surrounding area.

To move or extend parts of an image with the Content-Aware Move tool:

1. In the Tools panel, click and hold the Spot Healing Brush tool (🖉) to reveal the Content-Aware Move tool (✗) and select it.

2. On the Options bar, choose Move or Extend for Mode.

3. Adjust the Structure and Color settings to determine how much detail and color is preserved in the area around the object you're moving. Higher values blend the moved object more smoothly into its new surroundings.

4. (Optional) Turn on Transform On Drop on the Options bar, so that you will be able to rotate or resize content after you drag it to a new location.

5. Drag around the item you want to move or extend. Keep the selection loose so you don't lose parts of it when Photoshop blends it to its new location. (If you prefer, you can make a selection with any selection tool before selecting Content-Aware Move.)

6. Drag the selection to move or extend it.

7. Double-click inside the selection to apply the move or extension (**FIGURE 13.10**).

Using Replace Color

The Replace Color dialog enables you to target a range of colors using eyedropper tools and then adjust those colors with HSL (Hue, Saturation, Lightness) sliders or the Color Picker.

Replace Color works best when you want to change a difficult-to-select color throughout an image, but you can also use it to make a local color change in one area. For local changes, making a rough selection before using Replace Color is helpful. Replace Color cannot recolor neutral colors (white, black, and gray), however, because the changes it makes to hue and saturation are relative to the targeted color. (Instead, use a Hue Saturation adjustment layer with the Colorize option.)

To swap colors with Replace Color:

1. Make a selection around the area you want to recolor to prevent unwanted color changes elsewhere.

2. Choose Image > Adjustments > Replace Color. The dialog displays a preview of the selected colors in white, deselected colors in black, and partially selected colors in gray. (Press Ctrl/Command to see a preview of the image instead.)

3. (Optional) If the colors you want to change are limited to a contiguous area,

select Localized Color Clusters. This may give you more accurate results.

4. With the Eyedropper tool ( ), click the colors that you want to replace, either in the image or the preview box.

5. To select more or fewer colors, do any of the following:

   ▸ Shift-click with the Eyedropper or click with the Add To Sample Eyedropper tool ( ) to select more colors.

   ▸ Alt/Option-click with the Eyedropper or click with the Subtract From Sample Eyedropper tool ( ) to select fewer colors.

   ▸ Click the Selection Color swatch to open the Color Picker and choose the color you want to replace.

6. Increase or decrease the Fuzziness value by dragging the slider or entering a specific value to include more or fewer related colors.

7. Specify a Replacement color by changing the Hue, Saturation, and Lightness values (**FIGURE 13.11**) or clicking the Result swatch and selecting a replacement color with the Color Picker.

8. Click OK.

TIP If you have to perform the same color replacement in several images, you can use the Save and Load buttons in the dialog to reuse settings quickly.

FIGURE 13.11 Use the controls in the Replace Color dialog to modify colors you target by clicking with the eyedroppers or the Color Picker.

14

Smart Objects

Smart Objects let you nondestructively apply filters, transformations, and other edits to your images by preserving the original image data in a separate, linked file. So, for example, you can scale a Smart Object down, and later you can scale it back up without losing any image quality. When you want to make edits to the content of a Smart Object, it will open in a separate window in Photoshop, or in the original application (like Illustrator).

In Photoshop, you can create Smart Objects from regular layers, or use the Place commands to embed or link to other files. You can also add Smart Objects directly from Adobe Bridge, Lightroom, and Camera Raw. Linked Smart Objects update when changes are made to the source file. You can also duplicate a Smart Object so when you edit one instance, your changes are applied to all the others.

In This Chapter

Creating Embedded Smart Objects

You can make embedded Smart Objects by converting one or more layers in a Photoshop document or by placing a PSD, AI (Adobe Illustrator), TIFF, JPG, PNG, EPS, PDF, or camera raw file. Because they include all the data from the source image, however, placed embedded Smart Objects can significantly increase the file size of a Photoshop document.

Embedded Smart Objects are indicated by an icon (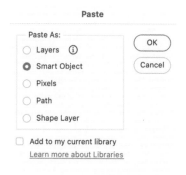) on the thumbnail in the Layers panel. You can also see that a selected layer is an embedded Smart Object in the Properties panel.

To convert one or more layers in a Photoshop document to an embedded Smart Object:

1. In the Layers panel, select one or more layers.

2. Choose Layer > Smart Objects > Convert To Smart Object. Or, right-click the selected layer name in the Layers panel, then choose Convert To Smart Object.

To add an embedded Smart Object from another file:

1. Select a layer; the Smart Object will be added above it.

2. Choose File > Place Embedded and select a file in one of the supported formats. For Illustrator AI and PDF files, the Open As Smart Object dialog opens where you can select a page, thumbnail size, and Crop To option.

3. Use the transform handles to size and position the new layer.

4. (Optional) Use the controls in the Layers panel to adjust the Blending Mode, Opacity, and Fill settings, which can be helpful when positioning and sizing the new layer.

5. To finish placing the Smart Object, double-click inside it on the canvas, then press Enter or click the Commit Transform button on the Options bar. (To cancel, press Esc or click the Cancel Transform button.)

TIP You can copy and paste objects from Illustrator or InDesign into Photoshop. InDesign content is automatically placed as a Vector Smart Object, while Illustrator content can be pasted as pixels, paths, layers, shape layers, or Smart Objects (**FIGURE 14.1**).

TIP You can also drag and drop a file from the Desktop (or any application that supports drag and drop, like Bridge or Lightroom Classic) into Photoshop to place it as an embedded Smart Object.

FIGURE 14.1 The dialog that appears when you paste Illustrator content into Photoshop includes the option to create a Smart Object.

To create an embedded Smart Object (instead of a Background layer) in a new file:

Do one of the following:

- In Photoshop, choose File > Open As Smart Object.

- In Adobe Camera Raw, click the down arrow on the Open Object button, then choose Open As Object from the menu.

TIP You can hold the Shift key to toggle the Open Image button to Open Object (FIGURE 14.2).

TIP You can edit Camera Raw preferences to open raw images in Photoshop as Smart Objects by default. Click the gear icon (✿) at the top right of the Adobe Camera Raw dialog to open Preferences. In Workflow preferences, under Photoshop, enable the Open In Photoshop As Smart Objects option.

FIGURE 14.2 Hold Shift to change the Camera Raw Open button to Open Object; click it to open the raw image as an embedded Smart Object in Photoshop.

Creating Linked Smart Objects

The primary benefit of linked Smart Objects (over embedded ones) is increased workflow efficiency. Because the linked image is separate and independent from the file where it's placed, a colleague could be working on a linked file while you work on the other layers in your Photoshop document. When your colleague saves their work, you can simply update the link to see their changes. In effect, you can both work on the same composition simultaneously.

Another benefit is reuse. You could have one file placed as a linked Smart Object in many different Photoshop files, and you'd only need to make a change in the one linked file to have that change appear everywhere. Because they include a flattened copy of the source image, linked Smart Objects don't increase the file size of a Photoshop document as much as embedded Smart Objects do.

On the downside, the links can be broken when a file is moved, renamed, or deleted. Plus, it's a more complicated way to work with more separate files to manage.

You can make linked Smart Objects by using the Place Linked command to place a file in one of the supported file formats or the Creative Cloud Library.

A link icon (⌘) appears on the thumbnail of a linked Smart Object in the Layers panel. You can also see that a selected layer is a linked Smart Object in the Properties panel and view the file path to the linked file (**FIGURE 14.3**).

To create a linked Smart Object from a file:

- Choose File > Place Linked.

To create a linked Smart Object from a Creative Cloud library:

1. Open the Libraries panel from the Window menu.

2. Choose a library.

3. Click a library item and drag it into the document window, or right-click the item and choose Placed Linked (**FIGURE 14.4**). Library-linked Smart Objects appear with a cloud icon (☁) on the layer thumbnail.

TIP You can also create a linked Smart Object by selecting a file in Adobe Bridge or Lightroom Classic and holding Option/Alt while dragging it into Photoshop.

FIGURE 14.3 You can tell the Grapes layer is a linked Smart Object by the layer thumbnail icon and the file path listed in the Properties panel.

FIGURE 14.4 Right-click a library item to place it as a linked Smart Object.

Managing Linked Smart Objects

Linked Smart Objects require a bit more monitoring and maintenance than embedded Smart Objects do. The main tasks you'll need to perform are updating links when the linked file is modified, fixing broken links, and relinking to a different file.

If the source file of a linked Smart Object is modified in another program while the containing Photoshop document where it was placed is open, the link will automatically update. However, if the containing Photoshop document is closed when the linked file is modified, the link will not automatically update. When you open the containing file, you will see an Out of Date icon ( ) on the layer thumbnail and you will need to manually update the link.

When the source file of a linked Smart Object is renamed, moved, or deleted, the link will be broken. A Relink dialog will appear when you open the containing document, showing the last known path to the linked file and a Relink button you can click to fix the broken link. If you cancel out of that dialog, you will see a Missing icon ( ) on the layer thumbnail. You must fix missing links before making any changes to the Smart Object or outputting your document.

To update a single Out of Date linked Smart Object:

Do one of the following:

- In the Layers panel, right-click a linked Smart Object layer and choose Update Modified Content.

- In the Properties panel, click the file path of the linked Smart Object and choose Update Modified Content (**FIGURE 14.5**).

To update all Out of Date linked Smart Objects at once:

- Choose Layer > Smart Objects > Update All Modified Content.

To fix a broken Smart Object link:

1. In the Layers panel, Layer > Smart Objects > Relink To File. Or, in the Properties panel, click the file path of the linked Smart Object and choose Relink To File.

2. Navigate to the file and click Place.

TIP You can prevent broken links by keeping source files for linked Smart Objects in the same folder as the containing document, because Photoshop will always check that folder when the containing document is opened.

TIP The same techniques for fixing broken links also can be used to relink to a different file.

FIGURE 14.5 Click the file path in the Properties panel to access useful commands for linked Smart Objects.

To relink to a different Creative Cloud library graphic:

1. In the Layers panel, right-click the layer and Choose Relink To Library Graphic. Or, in the Properties panel, click the file path of the linked Smart Object and choose Relink To Library Graphic.

2. From the Libraries panel that opens, select an item and click Relink (**FIGURE 14.6**).

VIDEO 14.1
Smart Object Workflow

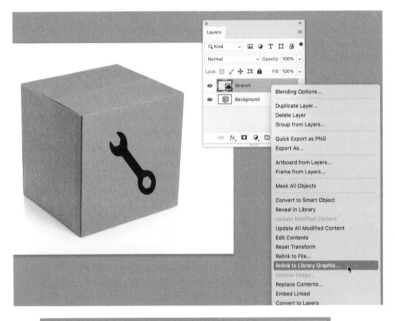

FIGURE 14.6 Relinking to a different library item can be a convenient way to collaborate with colleagues and access commonly used graphics in your workflow.

FIGURE 14.7 Double-click the thumbnail of a Smart Object layer to edit its contents.

Editing Smart Objects

When you want to make changes to the source content of a Smart Object, you need to open and edit it outside of the containing document. If the Smart Object was comprised of raster (pixel) data, it opens in a new window in Photoshop. If the Smart Object was comprised of content from another application, like Adobe Illustrator, it will open in that application. After you save the changes you made to the source content, they will be immediately reflected in the Smart Object. Note that unlike transforms and filters, these edits are destructive and cannot be reversed after the source file is saved and closed.

You can also replace the source contents of a Smart Object, in either a single instance or all instances. Importantly, any transforms or filters that were applied to the Smart Object are maintained when the source content is replaced.

To edit the contents of a Smart Object:

1. In the Layers panel, click the Smart Object.

2. In the Layers panel, double-click the Smart Object's thumbnail. Or, in the Properties panel, click Edit Contents.

3. Make your desired edits to the source file, then choose File > Save (**FIGURE 14.7**).

TIP When you're tweaking the contents of a Smart Object, it can be convenient to keep it open side by side with its containing file. To do this, choose Window > Arrange > Tile All Vertically. Each time you save the Smart Object file, the containing file will update automatically (**FIGURE 14.8**).

To replace the contents of a Smart Object:

1. Choose Layer > Smart Objects > Replace Contents.

2. In the dialog that opens, navigate to the file you want to use and click Place (**FIGURE 14.9**).

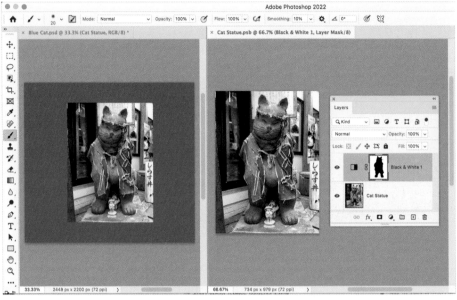

FIGURE 14.8 Tile windows to see the contents of a Smart Object and its container side by side.

FIGURE 14.9 When you replace the contents of a Smart Object, all Smart Filters, transformations, and effects are retained.

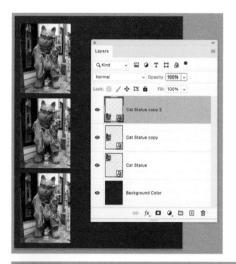

Duplicating Smart Objects

There are two methods for duplicating embedded Smart Objects, and which method you should choose depends on whether or not you want all the copies of the embedded Smart Object to change when you edit the source content from one instance.

Note that in both cases, the duplicate will be named the same as the original with a suffix of *copy*, and there is no other visual indication of the relationship between Smart Objects. So, you can't tell if a duplicate is linked to the original just by looking at them in the Layers panel. If that's important to you, you could manually rename the duplicate(s) or group them.

By contrast, all instances of Smart Objects that are linked to external files will change whenever you edit the source content from any of them.

To create a duplicate embedded Smart Object that is linked to the original:

1. In the Layers panel, click a Smart Object.

2. Choose Layer > New > Layer Via Copy. Or, drag a Smart Object layer to the Create New Layer icon in the Layers panel.

3. Any edits you make to the original Smart Object or the copy now will be reflected in the other (**FIGURE 14.10**).

TIP You can also drag the Smart Object layer to the Create A New Layer icon (⊞) in the Layers panel to create a linked duplicate.

FIGURE 14.10 When you duplicate an embedded Smart Object, you can edit any one of the copies and that change will be reflected in the others.

To create a duplicate embedded Smart Object that is independent from the original:

1. In the Layers panel, click a Smart Object.

2. Choose Layer > Smart Objects > New Smart Object Via Copy. Any edits you make to the original Smart Object or the copy will not be reflected in the other (**FIGURE 14.11**).

TIP You can also right-click the Smart Object in the Layers panel and choose New Smart Object Via Copy.

Exporting Smart Objects

You can reverse the process of embedding a separate file as a Smart Object by exporting it. Note that this is different from converting an embedded Smart Object to a linked one because the exported contents will not be linked to the file you export them from.

You can also use the Package command to create a folder with copies of all assets that comprise linked Smart Objects and the containing Photoshop document.

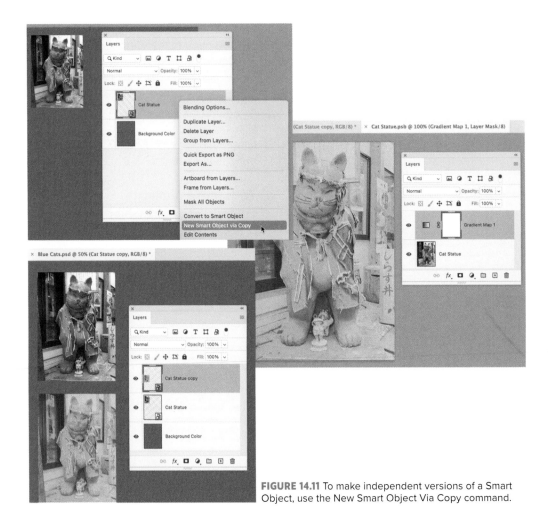

FIGURE 14.11 To make independent versions of a Smart Object, use the New Smart Object Via Copy command.

To export the contents of an embedded Smart Object:

1. In the Layers panel, click the embedded Smart Object, then choose Layer > Smart Objects > Export Contents.

2. In the Save dialog that appears, choose a location and filename for the contents of the Smart Object.

3. Click Save. If the embedded Smart Object was placed from another file, it will be saved in the same file format as the original. If it was created from layers, it will be saved in PSB format (**FIGURE 14.12**).

TIP You can also right-click an embedded Smart Object in the Layers panel and choose **Export Contents.**

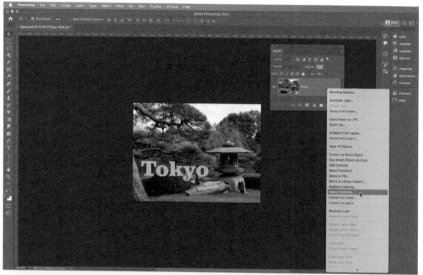

FIGURE 14.12
Exporting an embedded Smart Object comprised of layers gives you a PSB file comprised of those layers.

To package a Photoshop document along with its linked Smart Objects:

1. Choose File > Save.

2. Choose File > Package.

3. In the dialog that opens, choose a location for the package, which is a folder containing copies of the source files for all linked Smart Objects and the containing Photoshop document. The folder will be named the same as the Photoshop document (**FIGURE 14.13**).

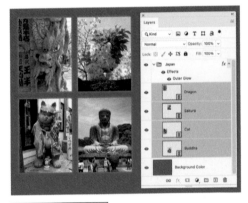

FIGURE 14.13 When you package a Photoshop file containing linked Smart Objects, you get a folder with a new copy of the containing file, plus a folder with copies of all the linked files.

Converting Smart Objects

Occasionally, you may want to convert a linked Smart Object to an embedded one, or vice versa. You can also convert Smart Objects to regular layers and rasterize them, making all transformations and filters permanent. Only do this when you're 100% sure you no longer need the ability to undo or modify the transformations, because once the file is saved and closed there's no going back.

To convert a linked Smart Object to an embedded Smart Object:

1. In the Layers panel, click a linked Smart Object.

2. Right-click the layer and choose Embed Linked, or click Embed in the Properties panel.

To update all embedded linked Smart Objects at once:

- Choose Layer > Smart Objects > Embed All Linked.

To convert an embedded Smart Object to a linked Smart Object:

1. In the Layers panel, click an embedded Smart Object.

2. In the Layers panel, right-click the layer and choose Convert To Linked, or click Convert To Linked in the Properties panel.

3. In the Save dialog that opens, choose a location and filename for the new linked source file. The new linked file will be saved in the same file format as the original placed file.

To convert a Smart Object to layers:

1. In the Layers panel, click a Smart Object.

2. In the Layers panel, right-click the layer and choose Convert To Layers, or click Convert To Layers in the Properties panel. A dialog will open.

3. If the Smart Object was comprised of a single layer, choose whether to retain the effects of transforms and Smart Filters or To discard them. Transforms and Smart Filters applied to Smart Objects comprised of multiple layers cannot be retained; the layers will be unpacked and put in a new layer group with the same name as the Smart Object (**FIGURE 14.14**).

To rasterize a Smart Object:

1. In the Layers panel, click a Smart Object.

2. Choose Layer > Smart Objects > Rasterize, or right-click the layer and choose Rasterize Layer.

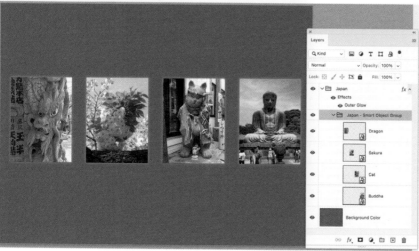

FIGURE 14.14 Unpacking a Smart Object to layers will remove any transformations but not effects.

Resetting Smart Object Transformations

If you've applied transformations to a Smart Object, such as scaling, skewing, rotation, or warping, you can remove those by clicking the Smart Object in the Layers panel and doing one of the following:

- Choose Layer > Smart Objects > Reset Transform.

- In the Layers panel, right-click the Smart Object layer and choose Reset Transform (**FIGURE 14.15**).

FIGURE 14.15 A Smart Object that has been transformed to fit into a scene can be reset to its original state without any loss of quality.

FIGURE 14.16 Use the Filter Type menu and buttons in the Layers panel to display specific types of Smart Objects.

Filtering the Layers Panel by Smart Objects

When working in a complex document with many layers it can be helpful to filter the display of the Layers panel to show Smart Objects according to their type and status.

To filter the Layers panel:

1. In the Layers panel, choose Smart Object from the Filter Type menu.

2. Click the buttons to display the kind(s) of Smart Objects you want to see (**FIGURE 14.16**), such as linked locally, linked from library, embedded, out of date, or missing.

3. Click the button to turn Layer Filtering on. Layers that do not match the type(s) of Smart Objects you selected will be hidden (**FIGURE 14.17**).

FIGURE 14.17 Filtering the Layers panel display makes it easy to spot missing linked Smart Objects.

Preferences Related to Smart Objects

The General preferences in Photoshop contain three settings related to Smart Objects.

- **Always Create Smart Objects When Placing:** If you turn this preference off, the Place Embedded command will place content as regular pixel layers instead of Smart Objects.

- **Resize Image During Place:** When this is turned on, placed files will be scaled to fit on the canvas, transformation handles will be displayed, and you need to accept or reject the transformation in the Options bar before continuing. If you turn this preference off, files will be placed at full size and no transformation handles will appear (**FIGURE 14.18**).

- **Skip Transform When Placing:** If you turn this preference on, placed files will be scaled to fit on the canvas and no transformation handles will appear.

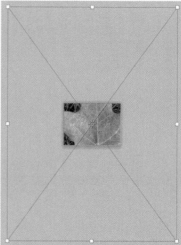

FIGURE 14.18 When the Resize Image During Place preference is turned on, placed Smart Objects will fit to the canvas. If you turn the preference off, new Smart Object layers will be placed at full size.

Essential Transformations

Beyond adjusting colors and retouching details, Photoshop enables you to transform layers, selections, and paths in ways ranging from the mundane (moving, scaling, rotating) to the miraculous (Puppet Warp, Content-Aware Scale). For example, in this chapter you'll learn how to change perspective, so it's as if the camera captured the image from a different angle. In addition to the basics of transformations, we've also included an obscure but powerful technique for increasing depth of field that uses Photoshop's ability to stretch and twist multiple layers into a perfectly aligned stack and blend them for maximum sharpness. So, whether you're wondering how to shift, skew, straighten, or spin, this chapter will transform your questions into answers.

In This Chapter

Using Free Transform

With the Free Transform command you can apply multiple transformations in a single operation, including rotate, scale, skew, distort, warp, and perspective. In fact, if you need to apply multiple transformations, it's a good idea to do them all in one operation because repeated transformations can decrease the image quality of pixel-based content.

You can apply most kinds of transformation to selections, individual layers, multiple selected layers, layer group, and paths. You cannot apply a warp to multiple layers at once, however. To achieve the same effect, convert the layers to a Smart Object first.

Transformations other than warps take place around a movable reference point that appears with the transform controls (**FIGURE 15.1**). You can choose the point from the Options bar (🔳) or set it manually by dragging or Alt/Option-clicking anywhere. You can even drag the reference point outside the boundaries of the canvas.

To apply Free Transform:

1. Select whatever you want to transform.

2. Choose Edit > Free Transform, or press Ctrl/Command+T. Transform controls appear around the item.

3. Follow the steps for any or all of the specific transforms.

4. To apply the transformation(s) to the item, press Enter/Return, click the Commit button (✓) in the Options bar, or move your pointer far enough away from the transform controls so it changes to a black arrow (⬆) and click.

TIP Just as you can apply multiple transformations in the same operation, you can also undo them one at a time by pressing Ctrl/Command+Z.

To move an item:

1. With the Free Transform tool, position your pointer inside the transform controls so it changes to a single black arrowhead (▶) and drag (**FIGURE 15.2**).

2. (Optional) To move an item a precise amount, change the X and/or Y values in the Options bar. Click the Relative Positioning button (△) to use relative values to reposition the item. For example, enter 100 px in the X value to move the item 100 pixels to the right.

FIGURE 15.1 The movable reference point for transformations

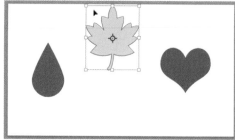

FIGURE 15.2 Moving an item by dragging

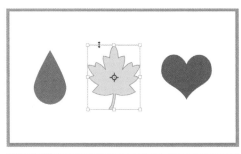

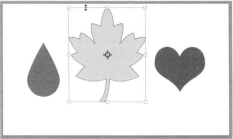

FIGURE 15.3 Scaling an item by dragging

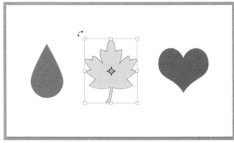

FIGURE 15.4 Rotating an item by dragging

To scale an item:

1. Position your pointer over the transform controls at any side or corner. When it changes to a double arrow (↔), drag (**FIGURE 15.3**). By default, the item will be scaled proportionally. Hold the Shift key as you drag to distort the item.

2. (Optional) To scale an item a precise amount, change the Width and/or Height values in the Options bar.

3. (Optional) Click the Maintain Aspect Ratio button (∞) to link the Width and Height values, so changing one will automatically change the other to maintain the item's proportions.

TIP You can also edit transformation values in the Transform section in the Properties panel.

To rotate an item:

1. Position your pointer just outside the corner transform controls so it changes to a curved double arrow (↰), and drag (**FIGURE 15.4**). Hold the Shift key to constrain the rotation in 15° increments.

2. (Optional) To rotate an item a precise amount, change the Rotation value in the Options bar.

To freely distort an item:

- Position your pointer over a transform handle and hold Ctrl/Command, so it changes to a white arrowhead (▷), and drag. Hold Alt/Option while dragging to distort relative to the center of the item (**FIGURE 15.5**).

To skew an item:

- Position your pointer over a side handle and hold Ctrl/Command+Shift, so it changes to a white arrowhead with a small double arrow (▷), and drag (**FIGURE 15.6**).

To apply perspective to an item:

- Position your pointer over a corner handle, hold Ctrl+Shift+Alt/Command+Shift+Option, so it changes to a white arrowhead (▷), and drag (**FIGURE 15.7**).

FIGURE 15.5 Freely distorting an item by dragging

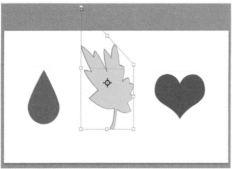

FIGURE 15.6 Skewing an item by dragging

FIGURE 15.7 Applying perspective to an item by dragging

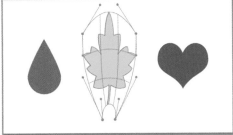

FIGURE 15.8 Warping an item by dragging mesh points or handles

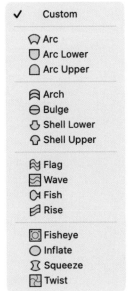

FIGURE 15.9 To warp an image, apply one of the 15 predefined warp options, or choose Custom to warp freely.

TIP The X, Y, Width, Height, Rotation, and Skew labels in the Options bar are all scrubby sliders, so you can simply position your pointer over them and drag to change values. To make finer adjustments, hold Alt/Option as you drag.

To warp an item:

1. Click the Switch Between Free Transform And Warp Modes button (🕸) in the Options bar. A warp mesh connected by control points appears on the item.

2. Drag the points and/or handles to warp the item (**FIGURE 15.8**).

3. (Optional) Choose a predefined warp style from the Warp menu in the Options bar (**FIGURE 15.9**).

4. (Optional) Adjust the effect of the predefined warp style by dragging the square white handle and/or using the controls in the Options bar to switch the warp orientation (🕸), bend, and distortion.

TIP If the warp mesh is hard to see because of the colors in your image, use the controls to change their display options.

To cancel all transformations:

- Press Esc, or click the Cancel button (⊘) in the Options bar.

Interpolation Methods

When you're transforming an item with any method except for Warp, you'll see an Interpolation menu in the Options bar. For the details on the different methods, see Chapter 3.

Using Perspective Warp

When working with photos of buildings, billboards, and other large objects, you may want to adjust the perspective (**FIGURE 15.10**). For this job, Photoshop provides the Perspective Warp command.

With Perspective Warp, you can draw four-sided shapes called *quads* to define planes in your image, and then straighten them so their sides are perfectly horizontal or vertical.

Perspective Warp requires that a compatible graphics processor is enabled in Preferences: Choose Edit > Preferences > Performance (Windows) or Photoshop > Preferences > Performance (macOS), and then confirm that Use Graphics Processor is enabled in the Graphics Processor area of the dialog.

To change perspective in an image:

1. Choose Edit > Perspective Warp.

2. In Layout mode, drag to create a rectangular quad.

3. Drag the corners (pins) or sides of the quad so they match the sides of the object whose perspective you want to change (**FIGURE 15.11**).

FIGURE 15.10 We'd like to fix this image so it looks like we're viewing this café straight on instead of looking up and slightly askew.

FIGURE 15.11 Drag the pins or sides of the quad as close as possible to the lines you want to make horizontal or vertical.

FIGURE 15.12 You can tell you're in Warp mode by three things: the Warp button in the Options bar is pressed, the pins turn black, and the grid lines inside the quad disappear.

4. Switch to Warp mode by clicking the button in the Options bar (**FIGURE 15.12**).

5. Drag the pins on the quad to change the perspective in the image. You can also use the buttons in the Options bar to automatically straighten near vertical lines, level near horizontal lines, or both (||| ≡ #) (**FIGURE 15.13**).

6. To accept the result, Press Enter or click the Commit Perspective Warp button (✓) (**FIGURE 15.14**). Or, to cancel and leave the image unchanged, press Esc or click the Cancel Perspective Warp button (⊘).

TIP You can create multiple planes by dragging out and positioning additional quads. The corners of the quads will snap together when you drag them close. This is useful when you want to make a subject such as a building or a car match the perspective of a background layer.

FIGURE 15.14 The finished image with corrected perspective. Note the resulting transparent areas in the corners, which will need to be filled or cropped out.

FIGURE 15.13 The straightened image in Warp mode

TIP In Layout mode, Shift-drag the side of a quad to constrain the shape of a plane while lengthening or shortening it.

TIP The document grid can help you see when lines in your image are horizontal or vertical. Turn it on and off by choosing View > Show > Grid or pressing Ctrl/Command+' **(FIGURE 15.15)**.

TIP In Warp mode, you can Shift-click the side of a quad to straighten and lock it in that orientation. The quad will appear in yellow **(FIGURE 15.16)**. To unlock it, Shift-click it again.

Keyboard Shortcuts

Try these handy keyboard shortcuts when using Perspective Warp.

* **W** switches to Warp mode.

* **H** shows and hides the edges of the quad in Warp mode.

* **L** switches back to Layout mode where you can reshape the quad without distorting the image.

FIGURE 15.15 A quick peek at the document grid will help you determine if you need to adjust the quad.

FIGURE 15.16 Shift-clicking a side of the quad locks it in a horizontal or vertical orientation.

Using Puppet Warp

Much as we wish all the images we work with were perfectly composed, occasionally things are just out of place and need to be moved. It could be something as small as an eyebrow or as massive as a skyscraper. Whether you need to change a facial expression or rearrange a city skyline, consider using Puppet Warp. It gives you the ability to move, stretch, and twist specific parts of an image by placing and dragging pins connected by a visual mesh. When used to maximum advantage, it's like having superpowers to reach into a photo and resculpt elements as if they were made out of clay.

While Puppet Warp can be used to create amazing special effects, you'll generally need to stick to subtle, small changes if you want to keep things looking realistic.

Puppet Warp works best on subjects that have been isolated on their own layer. Use Content-Aware Fill to fill in holes in the background, if necessary. Areas separated by transparency on the layer will have independent meshes.

To manipulate an image with Puppet Warp:

1. On the Layers panel, select the layer you want to manipulate.

2. Choose Edit > Puppet Warp.

3. Set pins in the image by clicking where you want to pull or rotate elements, as well as where you want to lock other elements in place (**FIGURE 15.17**).

FIGURE 15.17 Here, we've placed eight pins. One on the flamingo's head so we can move it, and the rest around its body to lock down the rest of the bird.

4. Drag pins to warp the image (**FIGURE 15.18**).

5. To accept the result, Press Enter or click the Commit Transform button (✓). Or, to cancel and leave the image unchanged, press Esc or click the Cancel Puppet Warp button (⊘).

While a nod may be as good as a wink in some cases, other times you need to refine your warp beyond, for example, our initial head movement. Use the following controls in the Options bar to change the effect of Puppet Warp:

- **Mode** controls the elasticity of the mesh. In most cases, the default Normal works fine. Choose Rigid to minimize pixel stretching and perspective effects, or choose Distort to maximize them (**FIGURE 15.19**).

- **Rotate** can be set to Auto or Fixed. Leave it on Auto if you want a mesh point to rotate automatically as you drag it. Set it to Fixed if you want to apply a specific rotation angle manually.

FIGURE 15.18 Dragging the pin on the flamingo's head moves and rotates it. Puppet Warp is so perfectly suited to manipulating flamingos we almost think it should be renamed Flamingo Warp!

FIGURE 15.19 The different results you can get with a pin in the same location, using the Puppet Warp Mode options: Normal, Rigid, and Distort.

FIGURE 15.20 By changing the depth of the pin on the flamingo's head, we can make it appear in front of or behind the rest of the bird.

FIGURE 15.21 Position your pointer next to a pin (not over it) and drag to rotate the pin.

- **Pin Depth** determines what happens when parts of the layer you're distorting overlap each other. Each time you click to set a pin, the new one is set above all others. Click the buttons in the Options bar (+ᴇ +ᴇ) to set pins forward or back in the stacking order and change the overlap effect (**FIGURE 15.20**).

- **Show Mesh** reveals the mesh applied to the layer.

- **Density** controls the spacing between mesh points. More points create a finer mesh and more precise results, at the cost of slower processing. Fewer points give you faster, less precise results.

- **Expansion** controls the overall size of the mesh. It gives you another way to control the spacing between mesh points. Positive values expand the mesh outward. Negative values will contract the mesh, hiding pixels at the edges.

TIP To rotate a pin without moving it, position your pointer adjacent to the pin, hold Alt/Option, and drag (**FIGURE 15.21**).

TIP Shift-click to select multiple pins so you can move or rotate them at the same time.

TIP To remove a pin, hold Option/Alt and click it.

TIP To remove all pins and start over, click the Remove All Pins (↻) button in the Options bar.

▶ **VIDEO 15.1**
Using Puppet Warp

Using Content-Aware Scale

Did you ever need to change the aspect ratio of an image, such as to recompose it for a widescreen video frame or square social media frame, but you didn't want to crop it because doing so would cut out important details? Content-Aware Scale can help in this situation. It allows you to exclude certain areas of your image from the transformation, leaving them untouched while you scale the rest.

The key to getting the best results with Content-Aware Scale is selecting what you want to prevent from scaling and saving that selection as an alpha channel that Photoshop can use during the transformation. Technically, you don't have to do this, but it helps so much that we consider it an essential step.

Note that Content-Aware Scale works on individual pixel layers and selections only. You can't apply it to multiple layers, Smart Objects, or masks unless you rasterize them first.

To use Content-Aware Scale:

1. Start by making a selection of the area you want to protect from being distorted when the image is scaled (**FIGURE 15.22**).

2. On the Channels panel, click the button to save the selection as a new channel (▫) (**FIGURE 15.23**).

FIGURE 15.22 Use any selection method to isolate the part of the image you want to keep unscaled.

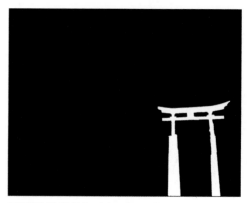

FIGURE 15.23 This alpha channel can be used to prevent the torii gate in the image (and its reflection in the water) from being affected when Content-Aware Scale is applied.

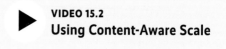

VIDEO 15.2
Using Content-Aware Scale

3. Deselect the selcted area.

4. (Optional) If you're working on the Background layer, Select All (Ctrl/Command+A).

5. Choose Edit > Content-Aware Scale.

6. In the Options bar, choose the channel you saved from your selection in the Protect menu.

7. Drag a handle on the bounding box. Hold Shift as you drag for non-proportional scaling (**FIGURE 15.24**). Alternatively, you can change the values in the Width and Height fields in the Options bar.

8. To accept the result, Press Enter/Return or click the Commit Transform button.

TIP To set a stationary point around which the layer is scaled, click a square on the reference point locator ( ) or drag the proxy ( ) on the canvas. This reference point has an effect only when you apply scaling via the Width and Height fields, not when you drag the bounding box.

TIP If the area protected by the alpha channel seems unnaturally separated from the scaled areas, you can allow it to scale by decreasing the Amount value with the slider in the Options bar.

TIP You can use the Protect Skin Tones button in the Options bar ( ) to have Photoshop attempt to recognize people and protect them from being distorted. However, you'll usually get much better results when you use the alpha channel method.

FIGURE 15.24 Dragging the bounding box scales the trees, sky, and water, but not the gate or its reflection.

Align and Blend Layers

Photoshop can transform multiple layers to align matching content in those layers. This is particularly useful when used in conjunction with another feature for blending the content in layers to maximize the sharpness throughout an image, increasing the depth of field. This technique is also known as *focus stacking*, and with it you can create images with a sharpness that is very difficult or impossible to achieve otherwise (**FIGURE 15.25**).

To merge multiple image files into one layered file, you can take advantage of automation features in Photoshop or Adobe Bridge. Note that when you use raw files in Adobe Bridge as your source, the current develop settings will be applied.

While it's not a requirement, you will usually get the best results if your source images are of stationary objects and were taken with a tripod.

FIGURE 15.25 The three source images. Note the limited depth of field and how a different avocado is in focus in each one.

To align and blend layers:

1. Do one of the following:

   - In Adobe Bridge, select the images you want to blend and choose Tools > Photoshop > Load Files into Photoshop Layers.

   - In Photoshop, choose File > Scripts > Load Files Into Stack. Use the Load Layers dialog to navigate to the files you want and choose them.

A new Photoshop file will be created containing separate layers with the name and content from each source image (**FIGURE 15.26**).

2. On the Layers panel, Shift-click all layers to select them.

3. Choose Edit > Auto-Align Layers. In the dialog, for Projection, choose Auto and click OK.

FIGURE 15.26 The new layered Photoshop file

4. Choose Edit > Auto-Blend Layers. In the dialog, for Blend Method, choose Stack Images and click OK.

5. Photoshop finds the areas of maximum sharpness and applies layer masks to reveal those areas in each source layer (**FIGURE 15.27**). Additionally, Photoshop creates a new merged layer combining the unmasked areas of the source layers on top of the layer stack (**FIGURE 15.28**).

FIGURE 15.27 Out-of-focus areas in each source image layer are masked out.

FIGURE 15.28 The topmost merged layer with increased depth of field

16

Filters

You can use filters to apply an incredible variety of modifications to layers or selections—from mundane (but essential) sharpening and retouching to creative effects that mimic anything from oil painting to halftones. Photoshop even offers a Neural Filters option that uses machine learning to help you change facial expressions, upsample images without losing quality, and colorize black-and-white images. You can apply multiple filters to the same layer, and alter the effects with blend modes, textures, masks, and patterns.

While all Photoshop filters can be applied to 8-bit RGB images, support for filters is more limited in CMYK images and images with higher bit depth. Filters that work only in RGB include Camera Raw, Lens Correction, Lens Blur, Oil Paint, and all the filters in the Filter Gallery.

In This Chapter

Applying Filters

Although you can target a regular layer or make a selection and immediately apply a filter, in most cases you shouldn't. Applying a filter in this way makes the change permanent. After you save and close the file, you can't revert to how it looked before you applied the filter.

A better option is to work nondestructively by converting the layer to a Smart Object and then applying the filter (called a Smart Filter). When you do this, Photoshop saves the filter as a layer effect that you can modify, disable, or remove at any time from the Layers panel. You can also move or copy Smart Filters between layers.

To apply a Smart Filter:

1. Click a layer in the Layers panel, or make a selection to limit the filter's effect to a specific area of the layer.

2. Right-click to the right of the layer name and choose Convert to Smart Object, or choose Filter > Convert for Smart Filters. Either way, an icon appears on the layer thumbnail indicating it has been converted to an embedded Smart Object (⊞).

3. Choose a filter from the Filter menu. Some filters will be applied immediately. Others, listed in the menu with an ellipsis after each name, will display a dialog where you can adjust settings and preview the effect before applying the filter.

Photoshop applies the filter as a layer effect, lists it under the layer name in the Layers panel under the heading Smart Filters (**FIGURE 16.1**), and automatically creates a mask. If you had an active selection before applying the filter, this mask will reveal only that area. Otherwise, it reveals the entire layer.

You can apply multiple filters to the same layer. In the Layers panel, they will all be listed under the Smart Filters heading and be affected by the same mask.

TIP A few exceptions to the Smart Objects rule: Lens Blur, Vanishing Point, Flame, Picture Frame, and Tree can be applied to pixel layers only.

VIDEO 16.1
Filters Overview

FIGURE 16.1 A filter applied to a Smart Object is known as a Smart Filter.

Modifying Smart Filters

After applying Smart Filters, you can toggle them off and on, modify their settings, change their Blending Options (opacity and blending mode), move or copy them to another layer, or delete them. Also, because filters are applied sequentially, when you apply multiple Smart Filters to one layer you can rearrange them to change their cumulative effect.

To change Smart Filter visibility:

In the Layers panel, do one of the following:

- Right-click to the right of the top-level Smart Filters entry, or the name of any individual Smart Filter and choose Disable/Enable Smart Filter.

- Click the Visibility icon to the left of the Smart Filters mask thumbnail, or any individual Smart Filter Visibility icon (👁).

- Click the Smart Filters mask thumbnail, then use any brush or painting tool to modify the mask. Painting with black on the mask hides the effects of the Smart Filters; painting with white reveals them (**FIGURE 16.2**).

TIP You can Shift-click a Smart Filter mask to disable/enable it and Alt/Option-click one to see the mask in the document window (instead of the full image), so it's easier to edit.

To change Smart Filter settings:

1. In the Layers panel, double-click to the right of the name of a Smart Filter to reopen the Smart Filter's dialog.

2. Tweak the settings as needed, and click OK. Multiple Smart Filters applied to one layer are applied in order from bottom to top, so if you modify any but the last (top) one, you will see an alert telling you that the other filters will not preview while the current filter is being edited.

TIP Naturally, filters that don't have a dialog (e.g., Average) can't be modified this way since there are no settings to change.

To change Smart Filter blending options:

1. In the Layers panel, double-click the Blending Options icon (≊) to the right of a filter name or right-click a filter name and choose Edit Smart Filter Blending Options from the context menu.

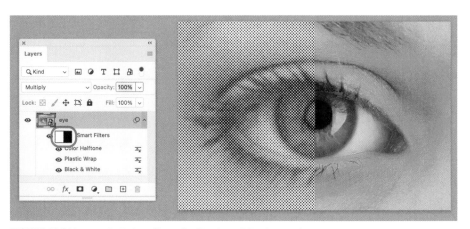

FIGURE 16.2 You can limit the effect of a filter by editing its mask.

2. In the Blending Options dialog that opens, change the Blending Mode setting and Opacity value for the Smart Filter (**FIGURE 16.3**).

3. Click OK when you're done.

To move a Smart Filter to another layer:

- Click and drag a filter effect name from one Smart Object layer to another (**FIGURE 16.4**).

To copy a Smart Filter to another layer:

- Hold Alt/Option, then click and drag a filter effect name from one Smart Object layer to another (**FIGURE 16.5**).

TIP You can apply multiple instances of a Smart Filter to the same layer by Alt/Option-dragging it within the stack of filter effects on that layer.

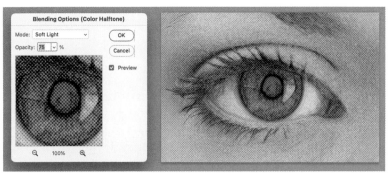

FIGURE 16.3 Apply blending modes and reduce opacity of a filter effect to blend it with other content in your image.

FIGURE 16.4 A simple drag and drop is all it takes to move a Smart Filter from one layer to another.

To remove one or more Smart Filters from a layer:

In the Layers panel, do one of the following:

- To remove all Smart Filters, right-click the top-level Smart Filters entry and choose Clear Smart Filters from the context menu. Or, drag the top-level Smart Filters entry to the Delete Layer button at the bottom of the panel (🗑).

- To remove a single Smart Filter, right-click a filter effect name and choose Delete Smart Filter from the context menu. Or, drag the filter effect name to the Delete Layer button at the bottom of the panel.

When multiple Smart Filters are applied to the same layer, they are listed in a stack in the Layers panel, consecutively from bottom (first) to top (last). Each filter is applied to the result of the ones below it. So, you can alter their cumulative effect by rearranging them in the stacking order.

To rearrange Smart Filters:

- Drag the filter effect name up or down in the stacking order (**FIGURE 16.6**).

FIGURE 16.5 Add the Alt/Option key as you drag a Smart Filter to copy it.

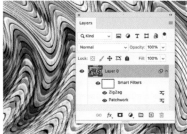

FIGURE 16.6 You get very different results depending on the order in which filters are applied. When Zig Zag is at the top of the list, it is applied last and distorts the regular grid pattern of the Patchwork filter.

Using the Filter Gallery

The Filter Gallery is a large dialog where you can preview and apply multiple creative effect filters at the same time. Only a subset of the filters listed in the Filter menu are available in the Filter Gallery—47 to be exact.

To apply filters via the Filter Gallery:

1. Choose Filter > Filter Gallery. The dialog that opens contains a large preview, a list of filter categories with thumbnails, and customizable settings (**FIGURE 16.7**).

2. Click a filter category name to display the available filters.

3. Click a filter thumbnail to preview its effects. The filter will appear in the applied filter list in the bottom-right corner of the dialog. You can toggle the preview on and off by clicking the Visibility icon (👁) next to the filter name in the list.

4. Modify the effect of the filter by changing the settings.

5. (Optional) Apply additional filters by clicking the New Effect Layer icon (⊞) and choosing them from the thumbnail list. Or, Alt/Option-click a thumbnail to add that filter (**FIGURE 16.8**).

6. (Optional) To remove an applied filter, click it in the list, and then click the Delete Effect Layer icon (🗑).

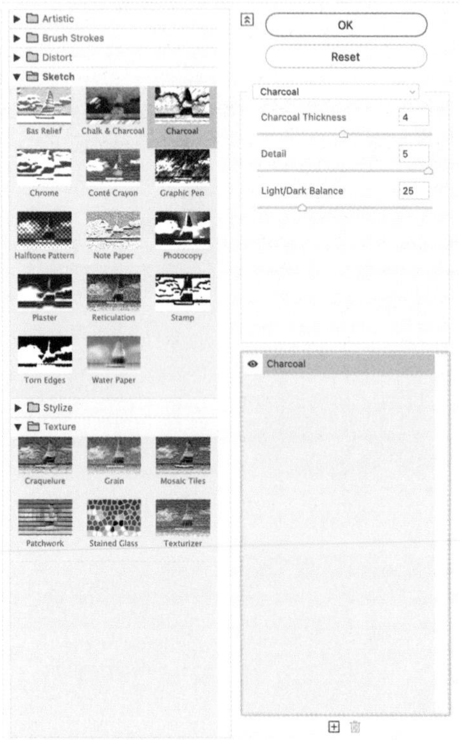

FIGURE 16.7 The Filter Gallery dialog

FIGURE 16.8 Build up complex effects by adding filters in the Filter Gallery dialog.

7. Click OK to apply the filter(s) and close the Filter Gallery dialog.

Just like in the Layers panel, each newly applied filter in the Filter Gallery will appear in the list on the right side of the dialog. Filters higher up in the list are applied to the result of the ones beneath. So, you can rearrange filters to change their cumulative effect by dragging a filter up or down in the list.

TIP You can toggle the visibility of the filter list and expand the preview area by clicking the double arrow to the upper right of the list (❮❮).

TIP You can choose a different filter at any time by picking one from the full list that appears below the OK and Cancel buttons at the top right of the dialog.

TIP You can apply multiple copies of the same filter to increase its effect. Modify the settings in each copy to create a complex and varied effect.

The Filter Gallery remembers the last-used settings and will apply them the next time you open the dialog.

To reset the Filter Gallery dialog to its default settings:

1. Hold Ctrl/Command. The Cancel button changes to Default.

2. Click Default to remove all filters from the applied list.

To apply the last used settings:

1. Hold Alt/Option to change the Cancel button to Reset.

2. Click Reset to reset the Filter Gallery dialog to the last used settings.

TIP To quickly see how different filters look applied to your image, click one thumbnail, then use your Up, Down, Left, and Right Arrow keys to navigate through the list of filters.

Sharpening Images

Let's be clear: Applying a sharpening filter in Photoshop cannot restore detail in a poorly focused image. It *can*, however, enhance the focus of existing details, making an image appear more crisp—a benefit to most images. Those destined for print especially benefit from sharpening. In fact, images optimally sharpened for print should appear slightly oversharpened onscreen to compensate for the blurring that occurs when ink spreads on paper.

The best filter for sharpening in Photoshop is Smart Sharpen. It offers more controls and options that can give you better results than its ancestor, the Unsharp Mask filter. For best results, apply it at the very end of your workflow, just before output, because sharpening has the potential to introduce undesirable artifacts that can be made more noticeable by subsequent transformations and adjustments.

Sharpening can be applied to one layer at a time only. So, if you want to apply nondestructive sharpening to an entire multilayered image, select all the layers and convert them to a single Smart Object. Then, you can apply sharpening to the Smart Object layer and still access the individual layers (in a separate document window) by double-clicking the Smart Object thumbnail in the Layers panel.

To sharpen an image:

1. Right-click a layer in the Layers panel and choose Convert To Smart Object.

2. Choose Filter > Sharpen > Smart Sharpen (**FIGURE 16.9**).

3. In the dialog, drag in the preview or click the canvas to preview the sharpening of a different area.

4. Choose an Amount value to set the strength of the sharpening. The optimal value will depend on the resolution and content of the image. Start at the default value and go upwards until the image looks oversharpened, then back down a bit (**FIGURE 16.10**).

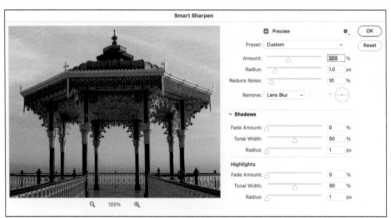

FIGURE 16.9 The Smart Sharpen dialog

FIGURE 16.10 Using an Amount of 400% results in noticeable oversharpening in the fine details.

5. Increase the Radius value until you see prominent halo effects in the image. Then, reduce the Radius to make the halos less noticeable (**FIGURE 16.11**).

6. In the Remove menu, make sure that Lens Blur is selected. Lens Blur is the only setting that automatically detects edges and will usually give you the best sharpening results.

7. Increase the Reduce Noise value to clean up any color noise introduced by the sharpening. In general, values between 10% and 20% work well.

8. (Optional) Expand the Shadows/Highlights category and use the controls to reduce oversharpening in the darkest and lightest areas of the image. Fade Amount is the most important of the settings here. The higher the Fade Amount the less sharpening will be applied.

9. Click OK to apply the sharpening and close the dialog.

10. (Optional) To adjust the blending mode or opacity of the sharpening effect, double-click the Blending Options button (⚏) on the Layers panel. To eliminate unwanted color shifts in sharpened areas, change the blending mode to Luminosity (**FIGURE 16.12**).

> **TIP** If the preview area in the Smart Sharpen dialog is too small for your tastes, click any corner of the dialog and drag to make it larger.

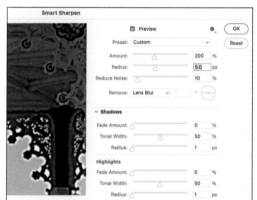 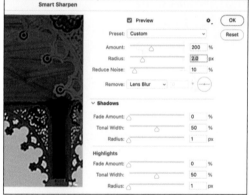

FIGURE 16.11 Adjust the Radius value until your image looks crisp but without obvious halos around edges. Here, 5 pixels is way to big; notice how it completely blackens some small elements. A Radius of 2 pixels is more appropriate.

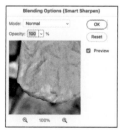 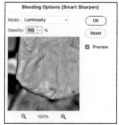

FIGURE 16.12 Changing the blending mode from Normal to Luminosity fixes the problem of glowing pink and green halos in this image.

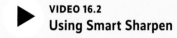
VIDEO 16.2
Using Smart Sharpen

Blurring Images

You can apply many creative effects in images by softening details through the use of blur features. Blur can also be used to reduce unwanted noise, grain, and textures.

Two of the most useful blur features are Motion Blur and Field Blur. Motion Blur simulates motion by blurring pixels by a specific distance and angle. Field Blur lets you create a gradient of blurs in an image by defining multiple blur points and adjusting the amount of blur for each one. With Field Blur you can direct the viewer's attention to certain elements by blurring others.

To simulate movement with Motion Blur:

1. In the Layers panel, click the layer you want to blur.

2. Choose Filter > Blur > Motion Blur.

3. In dialog that opens, set an angle and distance for the blur (**FIGURE 16.13**).

4. (Optional) Use the layer mask to limit the effect of motion blur to certain parts of the image.

5. Click OK.

Other Useful Blur Filters

Photoshop offers several other blur filters that you may find useful for certain tasks.

- **Lens Blur** attempts to simulate how a real lens creates depth of field, with features such as focal distance, iris options, and the ability to load a depth map from a camera or alpha channel.

- **Smart Blur** offers the most precise controls for blurring an image, using Radius and Threshold settings similar to Smart Sharpen.

- **Surface Blur** attempts to identify and preserve edges while blurring areas within those edges, which is useful for removing noise.

- **Tilt-Shift** can make photos of real-world settings seem like groups of miniature objects by blurring everything beyond a narrow area of focus.

- **Path Blur** enables you to add motion blurs along paths that you define.

- **Spin Blur** simulates circular motion by blurring pixels around one or more points from 0 to 360°.

FIGURE 16.13 To simulate the movement of the drummer's arm and drumstick, we added a Motion blur Smart Filter, then painted the layer mask to restrict the blur to the appropriate areas.

FIGURE 16.14 For a uniform effect across the image, we added four blur pins all with the default Blur value of 15 pixels.

FIGURE 16.15 The Blur values on the two sunflowers were reduced to 0 to bring them back into focus. A 7-pixel blur softens the stop sign slightly. A 15-pixel blur on the highway signs makes them totally out of focus.

To manipulate image focus with Field Blur:

1. In the Layers panel, click the layer you want to blur.

2. Choose Filter > Blur Gallery > Field Blur. Photoshop adds a blur pin in the center of the image. To move it, drag from its center.

3. Click elsewhere in the image where you want to define specific amounts of blur. Each time you click, another blur pin is added (**FIGURE 16.14**).

4. To adjust the amount of blur applied by a blur pin, click it and drag the blur handle (the circular slider around the pin). Or, change the value in the Blur slider in the Blur Tools area of the dialog (**FIGURE 16.15**).

5. (Optional) Use the other controls in the dialog to add bokeh effects, noise, or motion effects.

6. (Optional) Apply additional Blur Gallery effects by clicking the arrows to the left of the Blur Tools names and adjusting their settings (**FIGURE 16.16**).

7. Click OK.

TIP To remove a blur pin, click its center and press Backspace/Delete.

TIP To delete all pins and start over, click the Remove All Pins button ( ) on the Options bar.

FIGURE 16.16 Clicking an arrow enables one of the blur tools and reveals its settings.

Using the Camera Raw Filter

The Camera Raw filter shares many of the image-enhancement features found in the Adobe Camera Raw plug-in (see Chapter 20 online). With the Camera Raw filter, you edit a layer instead of an entire file. While the Camera Raw filter doesn't offer all the features of the plug-in, it's still quite powerful. With the Camera Raw filter, you can target adjustments to specific areas of an image without using layers or selections. You can quickly remove spots and unwanted elements without cloning or masking. You can also browse and choose from dozens of presets to instantly apply complex changes to images, including those in RGB, grayscale, and multichannel mode (but not CMYK).

To switch among the Camera Raw filter's five modes click the buttons in the top right of the dialog (**FIGURE 16.17**). Edit is the default mode. Spot Removal lets you eliminate distracting elements and small defects by painting over them or clicking. Masking, Red Eye, and Presets help you in the usual ways. The customizable presets also serve as handy tools for learning how various settings affect an image and offer creative inspiration (**FIGURE 16.18**).

FIGURE 16.17 The buttons to access the various modes of the Camera Raw filter (from top): Edit, Spot Removal, Masking, Red Eye, and Presets

FIGURE 16.18 Some of the many presets worth exploring in Camera Raw

To apply basic image adjustments with the Camera Raw filter:

1. In the Layers panel, click the layer you want to adjust.

2. Choose Filter > Camera Raw Filter. The Edit tools will be displayed.

3. (Optional) Click the Auto button. Camera Raw will use machine learning technology to analyze the image and apply what it thinks are the best settings to improve it. You can accept these settings or adjust them. Click the Auto button again to restore the default settings.

4. Click the arrow next to Basic to show all the options it contains.

5. Drag the sliders to apply adjustments. You can also enter specific values or use scrubby sliders by dragging over an adjustment's name (**FIGURE 16.19**).

6. To assess the effect of your adjustments, click the button below the preview area to cycle between various Before and After views (■). Or, click the button to toggle between the default and current settings ([▮]).

7. (Optional) To save your settings as a preset that you can apply with one click to any image, click the More Image Settings button (•••) and choose Create Preset.

8. Click OK.

TIP If you want to start over, hold Alt/Option and click the Reset button to return to the default settings. To reset only one adjustment or one category of adjustments, Alt/Option-click the adjustment or category name.

TIP Use the Zoom menu and the Fit button at the bottom of the dialog to adjust the zoom level of the preview.

TIP To apply Auto settings for the Basic adjustments one at a time, hold the Shift key and click the name of each adjustment.

Using Neural Filters

The Neural Filters use machine learning technology called Adobe Sensei to apply quick, nondestructive adjustments that would otherwise involve a complex and time-consuming manual process. Currently, Neural Filters can give you a leg up on tasks such as colorizing black-and-white photos, retouching portraits, and making two images match stylistically. Sometimes these filters yield underwhelming results, although a couple are quite useful. All of them are worth exploring and hold great promise as their development matures.

Specific Neural Filters are unavailable if Photoshop determines an image does not contain compatible content. For example, Skin Smoothing will be grayed out if Photoshop does not detect a face.

Some Neural Filters require an internet connection every time you use them, because they process image data in the cloud, not on your computer. As you apply settings, the progress bar that appears under the image preview indicates where the data is being processed.

When you invoke Neural Filters, Photoshop switches to a dedicated workspace with a simplified Toolbar and Options bar, and one large panel for applying Neural Filters. In the panel, you'll find the Neural Filters in two categories (**FIGURE 16.20**).

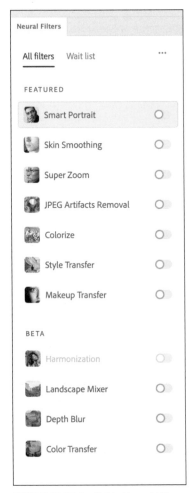

FIGURE 16.20 Available Neural Filters are grouped into Featured and Beta categories.

Featured filters the most mature and capable, while Beta filters are earlier in their development and may not produce good results. The Wait List area lists Neural Filters that are under development but not yet released, and it offers a button you can click to vote for a filter to be added in a future Photoshop update (**FIGURE 16.21**).

To apply a Neural Filter:

1. Click a layer in the Layers panel.

2. Choose Filter > Neural Filters.

3. If necessary, click the button to download the Neural Filter you want to use (⬆).

4. Click the Neural Filter's slider (◯) to enable it (●).

Generate unique photo realistic faces based on characteristics you specify.

I'm interested

FIGURE 16.21 The Wait List contains filters that are not yet ready for public release and may or may not eventually appear in Photoshop.

5. Use the controls on the right side of the dialog to apply the filter's effects (**FIGURE 16.22**).

6. (Optional) Apply additional Neural Filters by enabling them and adjusting their settings.

7. Choose an option from the Output menu. You can apply Neural Filter effects to the current layer, a new layer, a new document, or a Smart Filter.

8. Click OK.

TIP Click the Show Original button ( ) to compare before and after looks.

TIP Click the Layer Preview button ( ) to toggle the preview between Show Selected Layer and Show All Layers.

TIP If you want to start over, click the Reset Parameters button ( ) to reset all controls.

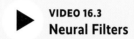

VIDEO 16.3
Neural Filters

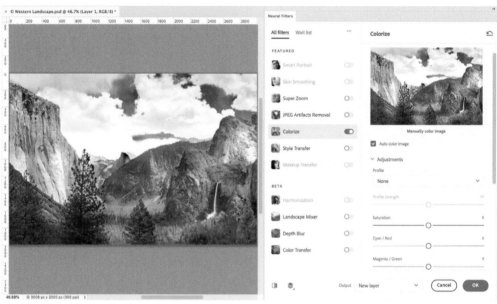

FIGURE 16.22 The Colorize Neural Filter may not produce perfect results, but it can give you a jump start and help you finish a job much faster.

Shape Layers and Paths

Did you know Photoshop also allows you to create graphics from vector paths, in the form of shape layers? These layers can be combined, modified, and formatted with strokes, fills, and effects. The advantage of using vectors instead of pixels is that they are resolution independent. You can output them at any size without a loss of quality. Edges and curves will always be crisp and clear.

You can create vectors with the Shape tools and the Pen tools. These tools require a bit more practice and effort to master, but in return they give you the ability to draw and edit any shape you want.

Vectors are also used in Photoshop for masking purposes. You can create and save vector clipping paths and vector masks to hide parts of an image.

In This Chapter

Using the Shape Tools

Photoshop offers six Shape tools: Rectangle, Ellipse, Triangle, Polygon, Line, and Custom Shape (**FIGURE 17.1**). Depending on the drawing mode you choose, you can use the Shape tools to add objects as live shape layers (which can have formatting like strokes, fills, and effects and appear in the Layers panel), unformatted vector paths (which appear in the Paths panel), or pixels.

To draw shapes:

Click and hold on one of the Shape tools in the Tools panel to reveal them all and select one. (Or, press Shift+U to switch between them.)

Select your desired drawing mode from the Options bar (**FIGURE 17.2**). Note that to draw in Pixels mode you must first select an existing pixel layer on the Layers panel where the new shape will be added.

1. Drag to draw the shape. As you drag you can see the width and height of your shape at your cursor. Hold the Shift key to constrain the width and height of the shape, so you can draw perfect squares and circles, or perfectly horizontal or vertical lines. Hold Option/Alt after you start dragging to draw from the center instead of the corner of the shape as you drag. Hold the spacebar after you start dragging to reposition the shape.

2. (Optional) If you used the Shape mode, adjust your shape at any time by selecting any Shape tool and using the controls on the Options bar, the Properties panel, or the shape itself (**FIGURE 17.3**). Note that the corner

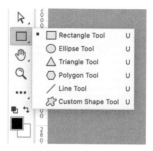

FIGURE 17.1 The Shape tools

FIGURE 17.2 The three modes for drawing shapes: Shape, Path, and Pixels

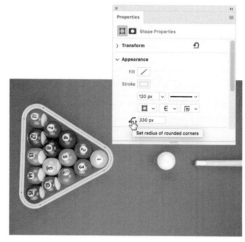

FIGURE 17.3 You can adjust the corner radius of a triangle shape in the Properties panel at any time.

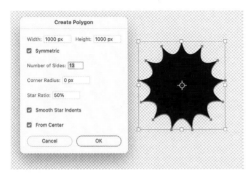

FIGURE 17.4 Clicking with the Polygon tool gives you a dialog where you can set various options.

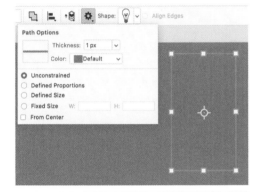

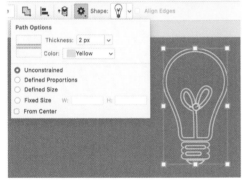

FIGURE 17.5 To see a path more clearly when it's over a similar colored background, select it and change its Color in Path Options.

radius shown in the Options bar is used only for new shapes that you draw. If you want to change the corner radius of existing shapes, use the Properties panel or drag the corner radius widget on the shape. Dragging a widget will change all the corners; hold Alt/Option as you drag to change just one corner.

TIP For more precision than dragging, click the canvas and set the options for your shape, like Width and Height. Note that some shape tools offer additional options (**FIGURE 17.4**).

TIP If a path is hard to see because it doesn't contrast with your image, click the gear icon on the Options bar and choose a different color and/or thickness (**FIGURE 17.5**). Note that this only changes the onscreen appearance of the path when it's selected.

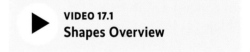

VIDEO 17.1
Shapes Overview

Formatting Shapes

Your options for formatting your shapes depend a lot on which mode you use when drawing. When you use Shape mode, for example, you can set the fill and stroke values at any time, before or after drawing the shape. You can fill shapes with solid colors, gradients, or patterns, or leave the fill transparent. Controls on the Options bar and Properties panel give you easy access to swatch groups, recently used swatches, and the Color Picker (**FIGURE 17.6**). If you draw with a Shape tool in Path mode, you cannot specify a stroke or fill, but you can convert the path to a live shape. If you draw in Pixels mode you can apply only a solid color fill and no stroke to your shape.

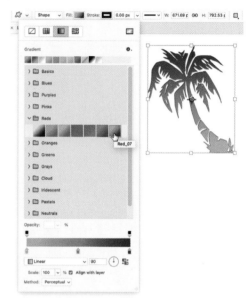

FIGURE 17.6 It's easy to apply any kind of fill to a live shape.

To convert a path to a shape for formatting:

1. Select the path in the Paths panel.

2. Select one of the Shape tools.

3. Click the Shape button on the Options bar.

4. Use the Appearance controls in the Properties panel to apply a fill and stroke as desired.

To choose the color fill for drawing in Pixels mode:

1. *Do one of the following:*

   ▸ Click the foreground color on the Tools panel and use the controls in the Color Picker.

   ▸ Use the controls in the Color panel.

2. Draw your shape.

Applying Dashed and Dotted Strokes

When you want to apply dashed or dotted strokes to your shapes, use the Stroke Options on the Options bar or Properties panel. Click one of the presets to apply the default dashed or dotted strokes (**FIGURE 17.7**).

If you want to customize the pattern, click More Options and set new Dash and Gap values (**FIGURE 17.8**). If you think you'll want to apply those same settings on a regular basis, click Save to create a custom preset.

You can also change the Align, Caps, and Corners settings for strokes here (**FIGURE 17.9**). Strokes can be aligned to the Center, Inside, or Outside of the path. Caps (the stroke at the ends of open paths) can be Butt, Round, or Square. Corners can be Miter, Round, or Bevel.

FIGURE 17.7 When you just need a quick dashed or dotted stroke, use the default presets.

FIGURE 17.8 Change the Dash and Gap values to create your own unique dashed or dotted stroke pattern.

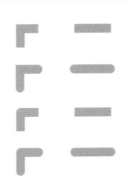

FIGURE 17.9 By default, strokes will be squared off, but you can achieve different effects by applying rounded caps, corners, or both.

Modifying Shapes and Paths

The basic shapes you can draw with the Shape tools are just a starting point. You can achieve complex and interesting effects by applying familiar transformations like rotating and scaling. After you draw a live shape or path with one of the Shape tools, you'll see transformation controls in the form of a bounding box with small squares on the corners and each side. You'll see these same controls any time you click on a shape with the Path Selection tool (▶). Manipulate these controls to modify your shapes.

You can also modify shapes by skewing them and moving one or more selected points. However, when you do so, the shape will be converted to a regular path.

To modify shapes:

Do one or more of the following:

- Drag one of the handles to change the height and/or width of the shape. Hold the Shift key as you drag to retain the original proportions of the shape. Hold the Alt/Option key as you drag to scale from the center.

- Position your pointer inside the shape and drag to move the shape.

- Position your pointer just outside the corner of the bounding box, so it changes to a curved double arrow (↱) and drag to rotate the shape. By default the rotation will happen around the center of the shape. To rotate around a different point on the canvas, drag the reference point proxy (⬦) on the canvas before rotating.

- To skew a shape, hold the Ctrl/Command key and drag a side handle.

- To free transform a shape, use the Direct Selection tool (▶) to select one or more points and drag them.

- To modify one or more points along the shape, use the Direct Selection tool (▶). Click to select individual points. Shift-click or drag over multiple points to select them all at once. Then drag to move the point(s).

TIP **If you need numerical precision when moving, rotating, or resizing a shape, use the Transform controls in the Properties panel.**

TIP **To delete a shape, click it in the Layers panel or on the canvas with the Path Selection tool, and then press Backspace/Delete.**

TIP **To switch between the Direct Selection tool and the Path Selection tool, hold the Ctrl/Command key and click a shape or path.**

Combining Shapes and Paths

You can combine basic shapes to create more complex shapes. By default, each time you draw a shape it will be placed on a new layer by itself. If you want to merge multiple existing shape layers into one, Shift-click them on the Layers panel, and choose Merge Shapes from the panel menu.

You can combine shapes on the same layer as you draw them using the Path Operations controls on the Options bar (**FIGURE 17.10**). After you draw shapes, you can combine them using the Path-finder commands in the Properties panel (**FIGURE 17.11**).

To combine shapes with Path Operations:

1. Using one of the Shape tools in Shape mode, draw your first shape.

2. Click the Path Operations button on the Options bar and choose Combine Shapes.

3. Draw your second shape. It will be added to the first shape layer. The two shapes will appear to be merged, but you can still select and manipulate them independently with the Path Selection tool () (**FIGURE 17.12**).

FIGURE 17.10 Path Operations controls you can choose on the Options bar when drawing shapes.

FIGURE 17.11 Pathfinder buttons in the Properties panel (from left): Combine Shapes, Subtract Front Shape, Intersect Shape Areas, Exclude Overlapping Shapes

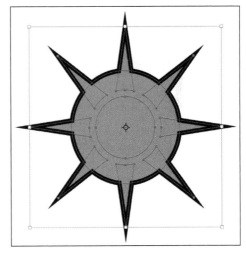

FIGURE 17.12 A sun shape from the Nature set of legacy shapes added to a circular shape drawn with the Ellipse tool

4. (Optional) Select the top path by clicking it with the Path Selection tool, and choose another Path Operation to create a different effect (**FIGURE 17.13**).

TIP You can achieve the same kinds of effects by combining shapes with Pathfinder commands in the Properties panel.

TIP When combining paths, you can use the Path Arrangement commands on the Options bar ( ) to move paths or shapes up or down relative to one another to change the effect.

TIP If Photoshop lags with a very complex shape layer, you may be able to speed things up by choosing Merge Shape Components from the Options bar. Note that doing so will convert the live shapes to regular paths but will not change their current appearance.

You can use alignment commands to precisely align or distribute paths and shapes relative to your current selection or to the canvas.

To align shapes on one layer:

1. Select one or more shapes with the Path Selection tool ( ).

2. Click the Path Alignment button on the Options bar ( ) and use the Align and/or Distribute commands. Note that you can align to the selection or to the canvas (**FIGURE 17.14**).

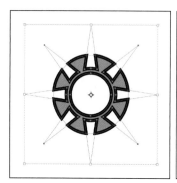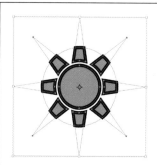

FIGURE 17.13 Different effects can be achieved by changing the way paths interact with each other.

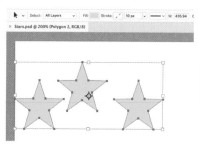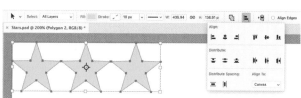

FIGURE 17.14 Three shapes on a single shape layer before and after distributing horizontal centers and top aligning relative to the canvas

To align shapes on multiple layers:

1. In the Layers panel, Shift-click to select the shape layers you want to align or distribute. Ctrl/Command-click if the layers aren't all next to each other in the stacking order.

2. Click the Move tool in the Tools panel.

3. Use the controls on the Options bar to align and/or distribute the paths or shapes relative to the selection or canvas (**FIGURE 17.15**).

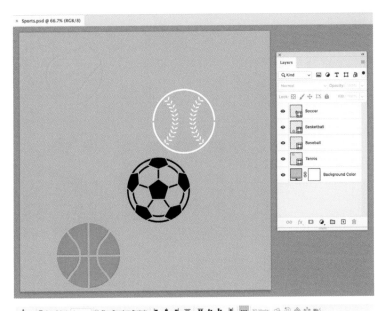

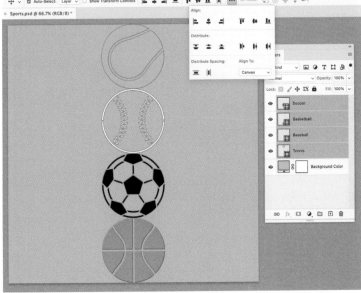

FIGURE 17.15 Four separate shape layers before and after center aligning and distributing vertical centers relative to the canvas

Using Custom Shapes

The Custom Shape tool is unique among the Shape tools because it gives you access to hundreds of complex vector shapes organized into groups such as flowers, people, buildings, and so on. You can also define new custom shapes from artwork that you draw or paste into Photoshop from Adobe Illustrator.

To add custom shapes:

1. Click the Custom Shape tool in the Tools panel (🐾).

2. On the Options bar, click the Shape menu and then choose a shape (**FIGURE 17.16**).

3. Click or drag on the canvas. Clicking gives you a dialog where you can set the precise width and height of your custom shape. To avoid distorting the shape, click Preserve Proportions in the dialog, or hold Shift as you drag.

4. (Optional) Change the Fill and Stroke on the Options bar or Properties panel.

TIP Click the gear icon in the Shapes menu to display custom shapes as thumbnails, list (thumbnail and name), or text only.

To define a new custom shape:

1. Draw vector artwork in Photoshop with the Pen tool in Shape mode. Or, copy vector artwork from Adobe Illustrator and paste it into Photoshop as a shape layer. Note that when you paste as a shape layer, the shape will lose its formatting from Illustrator and use the current foreground colors as its fill.

2. With one of the vector tools selected (Pen, selection, or shape), choose Edit > Define Custom Shape.

3. In the Shape Name dialog, give your new shape a name and click OK (**FIGURE 17.17**). The new custom shape will appear in the Shapes menu on the Options bar and in the Paths panel.

4. (Optional) Create a new shape group to house your shape or drag it into an existing shape group.

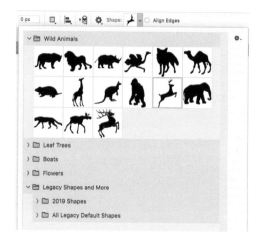

FIGURE 17.16 You can choose from hundreds of custom shapes that come with Photoshop.

FIGURE 17.17 You can save vector art that you draw or paste into Photoshop as a new custom shape.

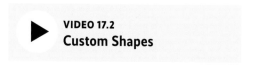

VIDEO 17.2
Custom Shapes

Using the Shapes Panel

Photoshop comes with dozens of premade shapes organized in groups that you can access from the Options bar and the Shapes panel when you're using the Custom Shape tool. The Shapes panel offers some key advantages, such as a search field where you can filter the display by typing the name of the shape you want in its Search field (**FIGURE 17.18**).

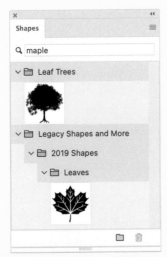

FIGURE 17.18 In the Search field, start typing the name of the shape you want to instantly narrow the display to matching items.

You also can display up to 15 recently used shapes at the top of the panel, just under the Search field. If you don't see it, choose Show Recents from the panel menu. To select one of the recent shapes, click it in the list.

The Shapes panel gives you the ability to add custom shapes to your document simply by dragging them from the panel onto the canvas. You don't even need to be using the Custom Shape tool.

Finally, the Shapes panel is the only place where you can fully manage shape groups. From its panel menu, you can create, delete, and rename groups.

Using the Pen Tools

With the Pen tools in Photoshop (**FIGURE 17.19**) you can draw new vector shapes and paths, as well as modify shapes that have been converted to regular paths or paths pasted from Adobe Illustrator. The Pen, Freeform Pen, and Curvature Pen enable you to draw new vector paths, while with the Add Anchor Point, Delete Anchor Point, and Convert Point tools you can modify points along an existing vector path.

With the Pen tool you create vector objects composed of straight or curved path segments connected by anchor points. Curves are controlled by handles attached to the anchor points. The length and angle of a handle defines the curve. The paths you draw with the Pen tool can be open (with endpoints that do not connect, like a V-shape) or closed (like a circle). While the Pen tool can be challenging for new users, it's worth the effort to master, because the precision and control it offers are unmatched by any other tool—making Pen tool proficiency a marketable skill.

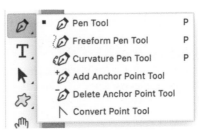

FIGURE 17.19 The pen tools in Photoshop

To draw with the Pen tool:

1. In the Tools panel, click the Pen tool (✐).

2. On the Options bar, select the mode you want: Shape to create a new shape with a fill and stroke or Path to create an unformatted vector path you can use for other purposes, like saving as a vector mask.

3. Do one or more of the following to draw your desired shape (**FIGURE 17.20**):

- Click to create an anchor point with no handles. This creates straight segments. Hold the Shift key to constrain the segments to 45° angles (a).

- Click and drag to create an anchor point with two handles. The handles move in unison to stay on opposite sides of the point, which define a smooth curve on both sides of the point (b).

- Click and, before moving your pointer, click and drag from the same spot. This creates an anchor point with one handle. The segment with no handle will be straight and the segment with the handle will be curved (c).

- Click and drag to define an anchor point with two handles. Then, move your pointer over the anchor point, hold Alt/Option, and click. This removes the handle for the next segment so you can go from a curve into a straight line (d).

- Click and drag to define an anchor point with two handles. Hold Alt/Option and click and drag on a handle to manipulate it separately from the other handle. This is how to define a curve coming out of a corner point (e).

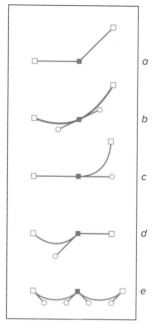

FIGURE 17.20 The five kinds of points you can draw with the Pen tool

The Freeform Pen Tool

There's not much to learn about using the Freeform Pen tool: Select it from the Tools panel (✐), then drag on the canvas. The path follows your pointer like you were drawing with a pen or pencil.

You can't beat the simplicity of using the Freeform Pen tool. We don't recommend you use it often, however, because the paths it creates are overly complex and imprecise. You can't control when new anchor points are created, so drawing with any kind or precision or efficiency is very difficult. The results are usually clumsy and harder to edit compared to the smooth curves you can get with the other Pen tools.

The Curvature Pen tool, a simpler alternative to the traditional Pen tool, enables you to draw vector paths simply by clicking and dragging points. There's no need to use other tools or keyboard shortcuts. If you get the hang of it, you may rarely need to use the traditional Pen tool.

To draw with the Curvature Pen tool:

1. In the Tools panel, click the Curvature Pen tool (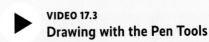).

2. On the Options bar, select the mode you want: Shape to create a new shape with a fill and stroke or Path to create an unformatted vector path you can use for other purposes, like saving as a vector mask.

3. Click to create the first anchor point.

4. Click again to set a second anchor point. This creates the first path segment. You can make it straight or curved. To create a curved segment, click once. To create a straight segment, double-click.

5. Drag the pointer to draw the next segment of your path. Release to complete the segment.

6. Continue drawing segments. Single-click for curves; double-click for straight lines. Click the initial anchor point to close the path.

To modify paths with the Curvature Pen tool:

Do one or more of the following:

- To adjust the shape of a curve, drag an anchor point. Notice that the adjoining path segments also change in response.

- Double-click an anchor point to toggle it between smooth (which creates curves) and corner (which creates straight lines).

- To move an anchor point, drag it.

- To add an anchor point, click anywhere on a path segment.

- To delete an anchor point, click it and press Backspace/Delete.

TIP **You can switch back and forth between any of the Pen tools while you are drawing a path to mix the abilities of each tool.**

▶ **VIDEO 17.3**
Drawing with the Pen Tools

Converting Paths into Selections and Masks

Along with channels, paths offer you a way to save selections for later use. Use paths only when you want to save a hard-edged selection. Use channels when you want a selection with softer transitions at the edges.

Saving a selection as a path also allows you to convert it into a custom shape or apply fills, strokes, and effects. Plus, paths can be readily converted into vector masks.

To convert a selection into a path:

1. Make a selection.

2. In the Paths panel, click the Make Work Path From Selection button (◇). A new path named Work Path will appear on the Paths panel.

3. Double-click Work Path in the Paths panel to rename and save the path (**FIGURE 17.21**).

To load a path as a selection:

1. In the Paths panel, click the path you want to make into a selection.

2. Click the Load Path As A Selection button (⚬).

To create a vector mask based on a path:

1. In the Paths panel, click the path you want to use as a vector mask.

2. Choose Layer > Vector Mask > Current Path.

FIGURE 17.21 Save a hard-edged selection as a path if you want to be able to reload the selection at any time.

18

Working with Type

To state the obvious: Photoshop is not a word processor. (It's called *Photo*shop, not *Word*shop). So don't make the mistake of trying to create documents with lots of text or complex layouts. Trust us, that's what Adobe InDesign is for. That said, Photoshop offers three methods for adding type: point text for single lines, paragraph text for multiline text, and type on a path for creating special effects. Plus, because Photoshop allows you to work in resolution-independent vector outlines for editable type, you can always output type at the full document resolution. So, there's no reason not to make ads, web graphics, posters, postcards, and maybe even a simple flyer in Photoshop. And you can apply unique creative effects to text that you can't do (or do easily) in other programs.

In This Chapter

Adding Point and Paragraph Text

Most of the time you'll be using point or paragraph text in your projects. Point text is best for short single lines of display type. Use paragraph text when you want long passages of type where lines automatically wrap.

Point text is left, center, or right aligned from the point where you click on the canvas and continues in a straight line (**FIGURE 18.1**). It will not wrap to another line unless you manually add a paragraph or other break character, so it can continue past the edge of the canvas.

To add point text:

1. With a Type tool, click the canvas.

2. Type or paste in the desired text. When you add type, Photoshop automatically adds a new type layer to the Layers panel.

3. Press Enter/Return to add line breaks as desired.

4. To accept the new text, click the Commit button on the Options bar. (Or, click Cancel if you've changed your mind.)

Paragraph text is contained within a bounding box that you set and modify with the Type tool (**FIGURE 18.2**). It wraps automatically to fit the box and will rewrap if you reshape the box.

FIGURE 18.1 Center-aligned point text. Notice the alignment point just after the letter *h*.

FIGURE 18.2 Paragraph text. Notice the bounding box with control handles surrounding the text.

Type Selection Methods

To edit text or change the formatting applied to it, you must first select a type layer (or a specific string of characters).

To select type for editing or formatting changes, do one of the following:

- On the Layers panel, click a type layer, select the Horizontal Type tool or Vertical Type tool (T or Shift+T), and then drag across text.

- To select one word: Double-click it.

- To select consecutive words: Double-click a word, then drag.

- To select a line of text: Triple-click the line.

- To select an entire paragraph: Quadruple-click anywhere in the paragraph.

- To select all the text in a type layer: Double-click the T icon on the Layers panel or quintuple-click (5 times) directly on type , or press Ctrl/Command+A.

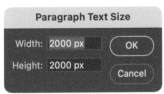

FIGURE 18.3
You can create a container for paragraph text with specific Height and Width values.

FIGURE 18.4 You can tell there's overset text in this paragraph by the plus sign in the bottom-right control handle.

FIGURE 18.5
Photoshop cannot convert overset paragraph text to point text, so it warns you to fix the problem (or accept the consequence) before converting.

TIP If you want to set a precise bounding box for paragraph text, Alt/Option-click, and enter the desired Width and Height in the Paragraph Text Size dialog (**FIGURE 18.3**).

TIP When you add a type layer, Photoshop will insert placeholder Lorem Ipsum text that you can type over or leave as is and replace later. If you don't want automatic placeholder text, go to Type Preferences and turn off Fill New Layers With Placeholder Text. At any time, you can fill the bounding box for paragraph type with placeholder text by choosing Type > Paste Lorem Ipsum.

To add paragraph text:

1. With a type tool, click and drag on the canvas to set a bounding box for the text.

2. Type or paste in the text. Photoshop automatically creates a new text layer in the Layers panel.

3. To accept the new text, click the Commit button on the Options bar. (Or, click Cancel if you've changed your mind.)

If you try to paste in more paragraph text than will fit in the box you created for it, part of the text becomes *overset*. It's still in the file, but it will not be visible.

To find and fix overset text:

1. Click the type with a Selection tool or a Type tool. If there's overset text, a plus sign appears in the corner handle of the bounding box (**FIGURE 18.4**).

2. Move, scale, or edit the text, or resize the box by dragging the control handles or changing the Width and Height values in the Properties panel.

If you change your mind after adding text, and want to convert from paragraph text to point text (or vice versa), the process is simple. However, if you are converting from paragraph text, you will need to fix any overset text first, as it will be deleted in the conversion (**FIGURE 18.5**).

To convert paragraph text to point text, or vice versa:

1. Click the T icon of the type layer on the Layers panel.

2. Choose Type > Convert To Point Text or Type > Convert To Paragraph Text.

Choosing a Font Family and Style

As someone once said, "fonts are the clothes words wear." Use these methods when it's time to dress up your text.

To change the font family and font style:

1. Select the type.

2. On the Options bar or Character panel, click the Font menu arrow to display a list of available font families and styles.

3. To preview how your text will look with a different font, move your pointer over the font in the list (**FIGURE 18.6**).

4. To apply the currently highlighted font, press Enter/Return or click the font name.

> **TIP** You can use the Up and Down Arrow keys to step through the font previews. To jump to the next font family, press Shift+Up Arrow or Shift+Down Arrow.

> **TIP** To change the size of the samples displayed on the Font menus, choose **Type > Font Preview Size**.

> **TIP** By default, Photoshop lists the 10 most recently used fonts at the top of the Font menus. You can change this number in Type Preferences to any value from 100 to 0 (to display no recent fonts). The change takes effect when you quit and relaunch Photoshop.

To search for specific fonts by name or style:

1. Select the type.

2. On the Options bar or Character panel, click in the Font Family field.

3. Start typing a font name or style, such as *semibold* or *condensed*. Photoshop will filter the font list to match what you type. You can also type partial names, such as *clar* to display every version of Clarendon active on your computer (**FIGURE 18.7**).

4. To preview a font in your document, move your pointer over the font in the list.

5. Press Enter/Return or click the font name to apply it to the text.

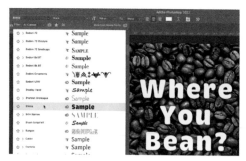

FIGURE 18.6 Move your pointer over a font in the menu list to see a preview of it applied to your text.

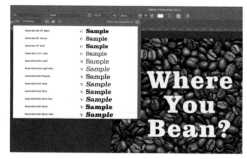

FIGURE 18.7 When you have an idea of which font you want to apply, start typing the font name or style in the Font menu, and Photoshop will filter the list to show only matching items.

FIGURE 18.8 Filtering the Font menu to show only a specific class of fonts can help you find the one you're looking for faster.

FIGURE 18.9 Use the three filtering buttons in the Font menu to browse fonts with specific characteristics.

TIP To display only fonts from a specific class of fonts (Serif, Sans Serif, Script, etc.) use the Filter menu at the top of the fonts menu. Choose All Classes to reset the menu to display all available fonts (**FIGURE 18.8**).

TIP You can also filter the Fonts menu to show only Adobe Fonts, fonts you have favorited, or Adobe Fonts that are visually similar to the current font (**FIGURE 18.9**).

TIP To change the formatting of all the text on a type layer, you don't have to select it first. Just click the layer in the Layers panel and use the controls in the Options bar or Character and Paragraph panels.

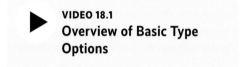

VIDEO 18.1
Overview of Basic Type Options

Applying Basic Text Formatting

Whether you're working with point text or paragraph text, the process is the same for applying basic formatting.

On the Options bar, you can do any of the following:

- Set the font family and font style (Regular, Bold, Italic, and so on).

- Choose or enter a font size (or drag the scrubby slider).

- Choose an anti-aliasing option to smooth the edges of text characters. In most cases, these options make only a very small difference in the appearance of type. Sharp creates the sharpest edges. Crisp is slightly less sharp. Strong makes text appear slightly heavier. Smooth softens the edge slightly. Mac LCD and Windows LCD mimic the way type is rendered by web browsers in those operating systems.

- Choose an alignment (Left, Center, or Right).

- Choose a color by clicking the Text Color swatch, and then using the Color Picker or Swatches panel. Alternatively, you can click anywhere on the canvas to sample a color.

TIP You can also apply colors, gradients, and patterns to live type layers via an editable Overlay layer style (see Chapter 22 online).

TIP To use the same formatting as an existing type layer, click that layer in the Layers panel before clicking or dragging on the canvas with the Type tool.

Kerning and Tracking

Kerning and tracking are typographic terms for changing the space between text characters. *Kerning* is used to change the spacing between a pair of text characters. *Tracking* is used to change the spacing across a range of characters, typically an entire paragraph.

To apply kerning:

1. With a Type tool, click between two characters on a type layer.

2. Do one of the following:

   ▸ On the Character panel, drag the Kerning icon (VA) left to decrease or right to increase the kerning. (**FIGURE 18.10**) (Or enter a positive or negative value in the kerning field.)

   ▸ Choose Metrics from the Kerning menu to apply the kerning value built into the current font (**FIGURE 18.11**).

   ▸ Choose Optical to let Photoshop determine the kerning based on the character shapes, which usually produces a tighter result (**FIGURE 18.12**).

To apply tracking:

1. Select the text with the Type tool or click the layer on the Layers panel to adjust tracking of all text on that layer.

2. On the Character panel, drag the Tracking icon (VA) left to decrease or right to increase the tracking. (**FIGURE 18.13**) (Or, enter a specific value in the associated field.)

3. To remove custom tracking, reset the tracking value of selected characters to 0.

TIP To make finer manual adjustments, hold the Alt/Option key while you drag the Kerning or Tracking icon.

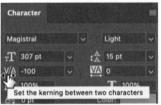

FIGURE 18.10 Drag the Kerning icon to quickly change the kerning between two characters.

FIGURE 18.11 The Metrics kerning method uses the default spacing set by the font designer.

FIGURE 18.12 Optical kerning can offer a quick way to tighten up a bit of type, with Photoshop choosing the optimal spacing between characters.

FIGURE 18.13 When you just want to eyeball the spacing in a bit of text, drag the Tracking icon.

FIGURE 18.14 The paragraphs of text are set with 70-point leading, which is slightly less than the default Auto leading value for 60-point type. The extra space between paragraphs was added in the Paragraph panel.

Justification			
	Minimum	Desired	Maximum
Word Spacing:	80%	100%	133%
Letter Spacing:	0%	0%	0%
Glyph Scaling:	100%	100%	100%
Auto Leading:	120%		

FIGURE 18.15 You can set the Auto Leading value in the Justification dialog box.

Adjusting Leading, Vertical Spacing, and Baseline Shift

Leading and baseline shift are the two features you can use to move horizontal type up and down. If you're working with vertical type, then tracking is the feature that will allow you to adjust the vertical spacing between characters.

Leading is the spacing between lines of type in a paragraph. The character with the highest leading value in a line of text determines the spacing of the entire line (**FIGURE 18.14**).

The Auto leading value is calculated as a percentage of the current font size. By default, it is set to 120%. So Auto leading applied to 20-point type will be 24 points. To change it from the default, open the Justification dialog from the Paragraph panel menu and enter a new Auto Leading value (**FIGURE 18.15**).

To adjust leading in horizontal paragraph type:

1. On the Layers panel, click a type layer.

2. On the Character panel, enter or choose a Leading value.

TIP If all the text on the layer has the same leading, you can change the leading by dragging the Leading icon right (increase) or left (decrease). Hold down Alt/Option as you drag to make finer adjustments.

Leading apples only to horizontal type. For vertical type, you change the vertical space between letters using tracking (**FIGURE 18.16**).

To adjust the spacing between characters in vertical type:

1. On the Layers panel, click a type layer containing vertical type.

2. On the Character panel, change the Tracking value.

To convert vertical type to horizontal type, or vice versa:

1. Double-click a type layer thumbnail.

2. On the Tool Options bar, click the Toggle Text Orientation icon (↓T).

You can use *Baseline Shift* to adjust the vertical position of type, typically raising or lowering characters relative to the others on the line (**FIGURE 18.17**).

To apply Baseline Shift:

1. Select the words or characters you want to shift.

2. On the Character panel, drag the Baseline Shift icon (A‡) right (raise characters) or left (lower characters), or enter a value. Positive values move characters upward from the baseline; negative values move them down.

TIP Hold down Alt/Option as you drag to make finer adjustments.

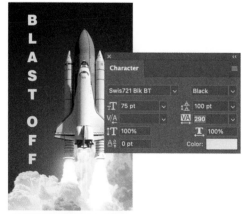

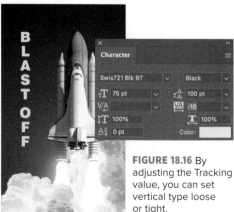

FIGURE 18.16 By adjusting the Tracking value, you can set vertical type loose or tight.

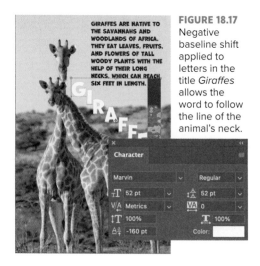

FIGURE 18.17 Negative baseline shift applied to letters in the title *Giraffes* allows the word to follow the line of the animal's neck.

Inserting Special Characters

You can achieve a more polished look with type by applying ligatures, fractions, and other alternate characters, provided you are using a font that contains those glyphs. Typically, OpenType "Pro" fonts offer the most special character options (**FIGURE 18.18**).

For example, you can replace a typed fraction like 1/2 with the typographically correct ½ glyph. You can also replace letter pairs such as *ff*, *ffl*, and *st* with ligatures that combine those letters into a single glyph (**FIGURE 18.19**). Or, if you are working with display type, you can insert swash and titling characters to add flair.

FIGURE 18.19 The craftsmanship in the image is echoed in the type through the use of standard ligatures (*Th* and *ft*) and a discretionary ligature (*st*).

To insert or specify alternate glyphs for OpenType characters:

1. With a Type tool, click in text to create an insertion point.

2. Open the Glyphs panel (Type > Panels > Glyphs), and scroll to find the glyph you want to insert.

3. (Optional) Use the Font Category menu to display a subset of the font, like Punctuation, Fractions, Dashes & Quotes, Symbols, and so on (**FIGURE 18.20**).

4. Double-click a glyph to insert it (**FIGURE 18.21**).

TIP The Glyphs panel keeps track of the 25 most recent glyphs you inserted. Double-click one of the recent glyphs to insert it again.

TIP To view and insert glyphs from a different font family and style, choose from the menus on the Glyphs panel.

FIGURE 18.20
The Font Category menu on the Glyphs panel gives you quick access to every subset of glyphs within a font.

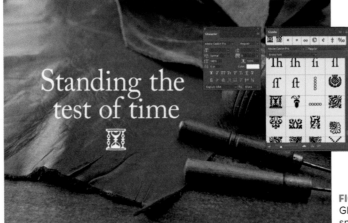

FIGURE 18.21 You can use the Glyphs panel to browse and insert special characters into text.

Formatting Paragraphs

You've formatted your text at the character level, applying fonts, colors, and so on. Now, it's time to take your design to the next level and apply formatting to paragraphs of text, setting the alignment, justification, indents, and spacing options. You can do all these things with the controls on the Paragraph panel.

To access paragraph settings:

1. With a Type tool, click in a paragraph or select a series of paragraphs. Or, if you want to apply settings to all the type in a layer, click the layer on the Layers panel.

2. If necessary, click the Toggle Character and Paragraph Panels button on the Options bar (▤). (Note that paragraph settings are also displayed in the Properties panel when you are working on a type layer.)

Alignment

You can use the three groups of controls at the top of the Paragraph panel to set the alignment of paragraph text: left, right, centered, or justified.

To choose an alignment:

1. Select the text you want to modify.

2. On the Paragraph panel, click Left Align Text, Center Text, or Right Align Text from the first group of icons (▤▤▤) to align type to the center or an edge of the bounding box (paragraph text) or initial insertion point (point text).

3. Click a button in the second group (▤▤▤) to justify text. These options make all lines but the last one span the full width of the bounding box surrounding paragraph text.

4. Click the last button, Justify All (▤), if you'd prefer to force all lines of paragraph text to span the full width of the bounding box.

About Paragraph and Character Styles

Photoshop offers the ability to save and apply settings to control the formatting of individual characters and paragraphs of text. You can access these by choosing Type > Panels > Character Styles Panel or Paragraph Styles Panel. However, we suggest that you ignore these Photoshop features, because they are incomplete, and encourage clumsy, inefficient workflows. If you have a large amount of text to work with, create your document in InDesign (where the text handling and style controls are much more robust) and place art from Photoshop into the InDesign layout.

If you just need to quickly duplicate the formatting in another type layer, simply duplicate that layer or copy and paste text from it and make any necessary edits.

Justification

You can use Justification settings to tighten or loosen the default spacing between words or letters in paragraphs of justified text. Justification settings can also apply glyph scaling to make characters wider or narrower. Taken together, these settings can have a strong impact on the overall flow and appearance of paragraph text. For example, use them when you need to fit more text in a particular area without changing the font or font size (**FIGURE 18.22**).

To change the spacing between words and letters in justified text:

1. Select the text you want to modify.

2. Choose Justification from the Paragraph panel menu.

3. In the Justification dialog, select the Preview option so you can see the effect of changing settings.

4. Use the controls in the Justification dialog to set Maximum, Desired, and Minimum values for Word Spacing, Letter Spacing, and Glyph Scaling.

Line wrap

Photoshop offers two paragraph composer options to control the way lines wrap in paragraph text. The Single-Line Composer handles each line separately. The Every-Line Composer takes every line in the paragraph into account to give a more balanced appearance to the "rag" end of the paragraph. When you edit text with the Every-Line Composer selected, all lines in the paragraph can rewrap.

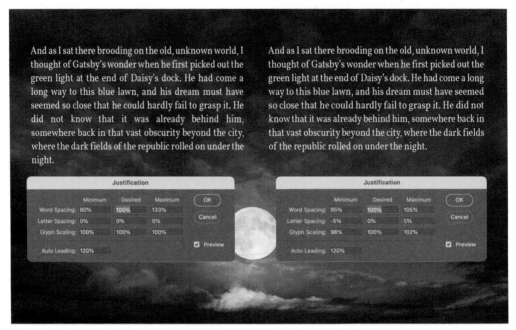

FIGURE 18.22 These two paragraphs have the same text and character formatting. The left paragraph uses default Justification settings. The one on the right has custom Justification settings that give the text a more consistent spacing throughout the paragraph, while using one less line. This is achieved by allowing small variations in letter spacing within words and glyph scaling, while allowing less variation in the spacing between words.

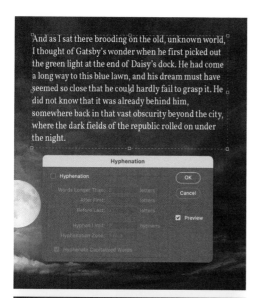

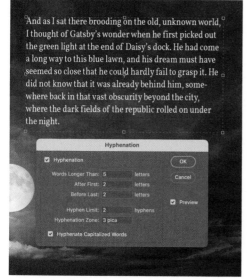

FIGURE 18.23 By allowing hyphenation you can often fit more text in a given space and achieve more consistent spacing.

To change the way lines wrap in paragraph text:

1. Select the text you want to modify.

2. From the Paragraph panel menu, choose either Single-Line Composer or Every-Line Composer.

To control hyphenation:

1. Select the text you want to modify.

2. Check or uncheck Hyphenate at the bottom of the Paragraph panel.

3. (Optional) To customize hyphenation settings, choose Hyphenation from the Paragraph panel menu and use the controls in the Hyphenation dialog (**FIGURE 18.23**).

Restoring Default Text Formatting

When you're experimenting with type formatting it can be easy to go too far and make a mess of things. When that happens, just choose Reset Paragraph from the Paragraph panel menu to return all currently selected type to the default paragraph settings.

Indents and margins

The Paragraph panel offers convenient controls for setting margins and indents on paragraphs of text.

To set indents:

1. Select the text you want to modify.

2. Use the controls on the Paragraph panel for Indent Left Margin, Indent First Line, or Indent Right Margin (**FIGURE 18.24**). Note that you can also use negative values to push text outside its bounding box.

To add space before or after a paragraph:

1. Select the text you want to modify.

2. Use the controls on the Paragraph panel for Add Space Before Paragraph or Add Space After Paragraph.

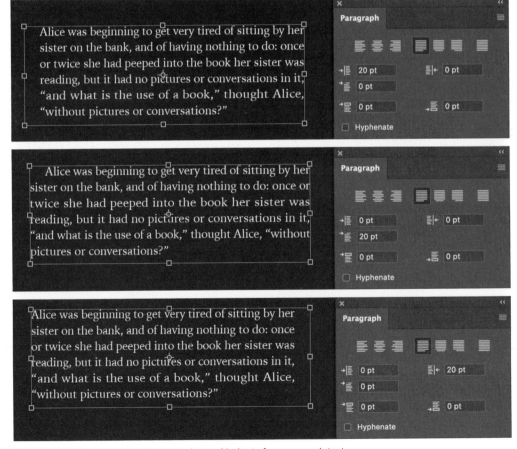

FIGURE 18.24 You can set various margins and indents for paragraph text.

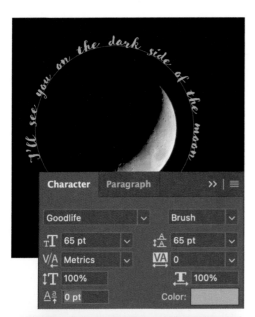

FIGURE 18.25 When you first add type on a path, the baseline is exactly on the path. To move it up or down, use Baseline Shift.

Working with Type on a Path

You can make type flow along the edge of an open or closed path created by a Pen or Shape tool. You can also use a closed path as a container for type.

To enter type along an existing path:

1. With a Type tool, position the pointer over the path so it changes its appearance ( ) and click.

2. Enter the type.

3. To adjust the vertical alignment of type on a path, use the Baseline Shift controls in the Character panel. Negative Baseline Shift values move the type down. Positive Baseline Shift values raise type above the path (**FIGURE 18.25**).

TIP Horizontal type is oriented perpendicular to the path. Vertical type is oriented parallel to the path. To change the orientation of type, select it and choose Type > Orientation > Horizontal Or Vertical.

To enter type inside a closed path:

1. With the Type tool, position the pointer inside the path so it changes its appearance (ⓘ) and click.

2. Enter the type. The path acts as a container, so lines automatically break wherever the type reaches the path boundaries (**FIGURE 18.26**).

You can change the position of type along a path and its orientation to the path by flipping it. When moving your type along a path, pay attention to the end of the type to make sure it all remains visible. If you move type so far that it will no longer fit on the path, you will create overset text. This text is still in the file, but it will not be visible until you fix the problem by moving, scaling, or editing the text. You can tell if there is overset text by clicking the path with a Selection tool or a Type tool. If there's overset text, a plus sign inside a small circle appears at the end of the visible text (⊕).

To move type along a path:

1. With either the Direct Selection tool or Path Selection tool, position your pointer over the type on a path. The pointer changes to an I-beam with an arrow (◀Ⅰ▶).

2. Click and drag the type along the path. Keep your pointer on the same side of the path; do not drag across it, or the type will flip (**FIGURE 18.27**).

TIP You can reposition type on a path using the Alignment controls on the Options bar or the Paragraph panel, making it left aligned, centered, or right aligned.

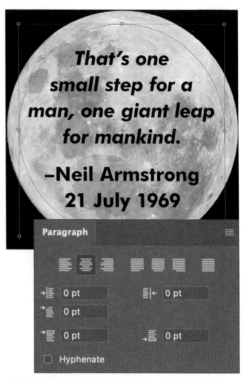

FIGURE 18.26 Center-aligned type set inside a circular path

FIGURE 18.27 By dragging with either Selection tool, you can move type along a path. If you drag across the path, the type will flip to the other side.

To move type by changing the shape of a path:

1. On the Layers panel, click a layer containing type on a path.

2. With a Type tool or Selection tool, click the path on the Paths panel to make it active.

3. To change the path's shape, do one of the following:

   ▸ Choose Edit > Free Transform Path and then reshape the path by dragging with the control handles or change the width, height, angle, or skew in the Options bar.

   ▸ With the Direct Selection tool, click and drag anchor points to move them.

4. The type moves to conform to the new shape of the path.(**FIGURE 18.28**).

TIP If you need more anchor points to achieve the shape you envision, use the Pen tool to add, remove, or convert anchor points to reshape the path.

To flip type to the other side of the path:

1. With either the Direct Selection tool or Path Selection tool, position your pointer over the type on a path. The pointer changes to an I-beam with an arrow.

2. Click and drag across the path.

TIP If you want to move type across a path without changing the direction of the type, use the Baseline Shift controls in the Character panel. Negative values move type down, positive values move type up.

Warping Type

You can create special effects by applying warp styles to type. Any warp style you apply becomes an attribute of the type layer, so you can make changes to it at any time.

To warp type:

1. Select a type layer.

2. With a Type tool, click the Warp button on the Options bar ().

3. Choose a warp style and orientation (Horizontal or Vertical).

4. (Optional) Specify values for such warping options as Bend, Horizontal Distortion, and Vertical Distortion (**FIGURE 18.29**).

5. Click OK.

To change or remove a warp effect:

1. Select a type layer.

2. With a Type tool, click the Warp button on the Options bar.

3. Apply different settings in the Warp Options dialog, as desired. Or, to remove a warp effect, choose None from the Style menu.

4. Click OK to apply the changes or remove the effect.

FIGURE 18.29 Use the Warp feature to reshape type with a variety of nondestructive effects.

Painting on a Type Layer

If you want to apply certain creative effects to type, like painting directly on type with the Brush tool or distorting the letters with the Smudge tool, you must convert it to pixels first via the Rasterize Type command (**FIGURE 18.30**).

Note that most other features that cannot be directly applied to live type, like filters, or the Transform > Distort or Perspective commands, *can* be applied to Smart Objects. So, it is not necessary to rasterize type to use these features. Instead, convert the type layers to Smart Objects. Then, if you need to modify the text, you can simply edit the Smart Object (see Chapter 14).

Also, in some cases, you may be able to create the effects you want by painting on a separate layer, while keeping your live type intact.

If you need to rasterize a type layer, it's a good idea to keep a backup of the original text in case you need to make changes to it. To preserve a copy of the editable type layer, click on it in the Layers panel and press Ctrl/Command+J.

To rasterize a type layer into pixels:

- Right-click the type layer name and choose Rasterize Type. The layer thumbnail changes from a T to the type shapes surrounded by transparent pixels.

FIGURE 18.30 Before you can create smoky letters like these with the Smudge tool, you have to rasterize the type layer.

Replacing Missing Fonts

The problem of missing fonts occurs when you open a document that uses fonts not installed on your computer.

On the Layers panel, a type layer with missing fonts appears with a yellow warning icon (⚠).

Photoshop will automatically activate missing Adobe Fonts when you open a document that uses them, provided you are online and logged in to your Creative Cloud account. While missing Adobe Fonts are syncing to your computer, you will see a blue download icon on the Layers panel. When the fonts are synced, the icon will disappear, and you can work with the type as normal.

If you activate missing fonts from a source other than Adobe while a document is open, you may see a gray warning icon on a type layer thumbnail (⚠). Double-click it to update the layer before proceeding. Or, if there are multiple layers with gray warning icons, choose Type > Update All Text Layers.

If you are unable to activate missing fonts automatically, you can replace the missing font manually, or use the Manage Missing Fonts feature to resolve the situation.

To manually resolve missing font issues:

1. On the Layers panel, double-click the thumbnail of the type layer with missing fonts.

2. In the dialog that appears, do one of the following:

   ▸ Click Replace and choose a different font in the Options bar.

   ▸ Click Manage, and use the options in the Manage Missing Fonts dialog to either replace the missing font with the default font or with a font already used in the document (**FIGURE 18.31**).

TIP You can quickly replace all missing fonts at once by selecting Replace All Missing Fonts With The Default Font

FIGURE 18.31 Double-click the type layer thumbnail to replace a missing font.

Matching Fonts

If you have an image containing type that is not live, you can use the Match Font feature to find visually similar fonts in the Adobe Fonts collection.

To find and activate matching fonts:

1. Select Type > Match Font.

2. Adjust the rectangular marquee so it encompasses up to three lines of type. The more different characters in the sample, the more accurate the search results will be.

   Photoshop displays a list of fonts that are similar to the text in your selection.

3. To activate an Adobe Font from the list, click the cloud icon (**FIGURE 18.32**).

When the Match Font feature works, it can be a great time saver. The problem is, it frequently doesn't work, which can be frustrating. If you're not satisfied with the initial search results, try readjusting the marquee to force Photoshop to search again. You may also get better results with other font matching services like What the Font.

FIGURE 18.32
The Match Font feature is a fast way to identify typefaces in raster images, and then download the exact font or one that's similar.

Creating a Text Sandwich

Here's another timeless text technique that almost every Photoshop user should have in their bag of tricks. You've seen it countless times on magazine covers where a model's head (or some object) overlaps the magazine title.

To make a text sandwich:

1. Identify which portion of the image you want to overlap the text and make a selection of it.

2. Copy that selection to a new layer (Ctrl/Command+J) and drag that layer above the type layer (**FIGURE 18.33**).

FIGURE 18.33
By copying a selection of the image to a separate layer above the type layer, you can create the text sandwich effect.

Filling Type with an Image

One common job that Photoshop users are asked to perform is filling type with an image. When it's done well, the shapes of the type outlines combine with the content of the image to convey a message that's greater than the sum of its parts. There are various ways of accomplishing this, but the most flexible and efficient is to use a clipping mask.

To fill type with an image:

1. Start with a file containing the image you want to put in type. If the image layer is the Background layer, click its lock icon on the Layers panel to convert it from a background layer into a regular layer.

2. Create a type layer with the text you want and format it as desired. Large, bold, sans serif type usually works best.

3. On the Layers panel, drag the image layer so that it is immediately above the type layer.

4. Create a clipping mask by Alt/Option-clicking on the divider between the image layer and the type layer on the Layers panel. The image appears inside the text (**FIGURE 18.34**).

TIP To adjust the position of the image within the text, select the Move tool, click the image layer, and drag the image.

TIP To move the text outlines in relation to the image, select the type layer and use the Move tool to move the text. Note that you may need to turn off Auto-Select in the Options bar to avoid selecting the image layer when you click and drag on the canvas.

FIGURE 18.34 A clipping mask is the simple key to putting an image in text. Best part: The text remains editable.

Printing and Exporting

As fun as it is to work with files in Photoshop, the entire reason for their existence, their *raison d'être*, is to be seen in some kind of output, either in print or onscreen.

You can, of course, directly print from Photoshop, but it's more likely you'll be handing files off to a print service provider or that your images will be placed into other documents, such as Adobe InDesign layouts, before output. To support professional print workflows, Photoshop offers features for soft proofing (simulating print output onscreen) and conversion to CMYK mode using color profiles.

You can use artboards to design images for viewing on multiple screen sizes and scenarios and export them in any desired size and file format. You can also make images that move by exporting them as animated GIFs.

If you need to hand off your files to the next person in the workflow, you can use the Package command to gather up all linked image assets in one convenient folder.

In This Chapter

Designing with Artboards

Artboards are a special kind of layer group you can use to organize and output content in various sizes. Any elements you place on an artboard are clipped to its boundaries, so in effect each artboard appears like a separate canvas within Photoshop. You can add layers, layer groups, and Smart Objects to an artboard. Artboards are optimized to work in RGB mode and are particularly useful for designing layouts for the web and mobile app interfaces, where you can benefit from keeping different compositions in one file and view them side by side. You can remove artboards from a document and output artboards to separate files.

To create a document with artboards:

1. Choose File > New.

2. In the New Document dialog, select the Artboards option. If you choose one of the default blank document presets from the Web or Mobile categories, artboards will be turned on automatically (**FIGURE 19.1**).

3. Set your other desired document options, and click Create.

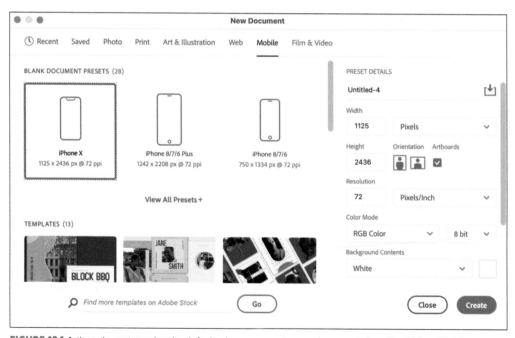

FIGURE 19.1 Artboards are turned on by default when you create new documents from the Web or Mobile presets.

To convert an existing document to an artboard document:

1. In the Layers panel, select one or more layers or layer groups.

2. Right-click and choose Artboard From Layers or Artboard From Group. If you are creating an artboard from layers a dialog appears where you can choose a size from a preset menu and name your new artboard (**FIGURE 19.2**). Note that you must first convert the Background layer to a regular image layer before you can convert it to an artboard.

To add more artboards to a document:

1. In the Tools panel, click and hold on the Move tool (✛) to reveal the Artboard tool (⌐) and select it.

2. Click and drag on the canvas to draw a new artboard. Or, in the Layers panel, click the name of an existing artboard, then click one of the plus icons (⊕) that appear around it to duplicate it. The new artboard will have the same properties as the existing one. Alt/Option-click to duplicate the contents

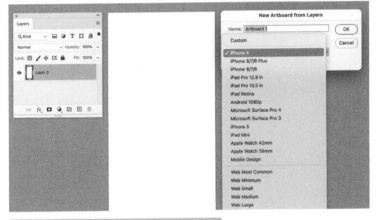

FIGURE 19.2 Before and after converting a regular image layer to an artboard

of the existing artboard to the new one (**FIGURE 19.3**).

3. (Optional) Use the Properties panel to change the size, position, preset, or background color of the artboard.

To rename an artboard:

- In the Layers panel, double-click an artboard name to edit it. If you don't see any artboard names, choose View > Show Artboard Names to reveal them.

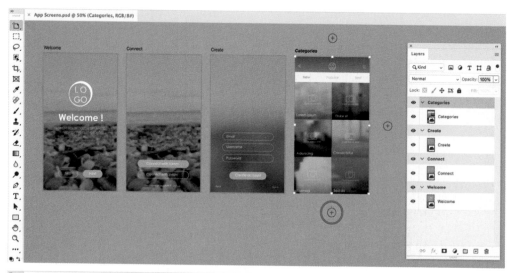

FIGURE 19.3 It's easy to duplicate an artboard with or without its contents.

FIGURE 19.4 Ungrouping an artboard leaves its contents in place while removing the artboard itself.

To reposition an artboard:

Do one of the following:

- With the Move tool or Artboard tool, click and drag the name of an artboard on the canvas.

- In the Layers panel, click an artboard and change the X and Y values in the Properties panel.

TIP You can align and distribute multiple artboards by selecting them in the Layers panel and choosing Layer > Align/Distribute, or using the Align controls on the Options bar.

To remove artboards:

Do one of the following:

- With the Move tool or Artboard tool, click the name of an artboard on the canvas and press Backspace/Delete.

- In the Layers panel, select one or more artboards and click the Delete Layer button. Photoshop will ask if you want to delete the artboard and its contents or the artboard only.

- To remove an artboard but leave its contents in place, select it in the Layers panel, then right-click and choose Ungroup Artboards, or press Ctrl+Shift+G/Command+Shift+G (**FIGURE 19.4**).

To move or copy elements between artboards:

Do one of the following:

- With the Move tool, drag the element from one artboard to another. Hold Alt/Option as you drag to duplicate the object instead of moving it.

- In the Layers panel, drag the element to the desired artboard. Alt/Option-drag to duplicate the object instead of moving it.

To save artboards as separate files:

1. Choose File > Export > Artboards To Files.

2. In the Artboards To Files dialog, specify where you'd like to save the file (Browse), whether to include overlapping areas or only the contents within each artboard, whether to export only the artboards currently selected in the Layers panel or all artboards, whether to include the artboard background colors, and the file type.

3. (Optional) Specify a prefix for the filenames. Leave this field blank to have the files named the same as the artboards are in the Layers panel.

4. Click Run. A confirmation dialog appears when the process is complete (**FIGURE 19.5**).

TIP By default, artboards appear with a gray border in Photoshop. If you want to remove this, go to Interface preferences and change the Artboards Border setting from Line to None.

TIP For whichever file type you choose to save your artboard as, you can also turn on and customize Export Options.

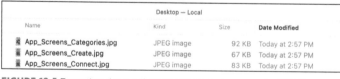

VIDEO 19.1
Working with Artboards

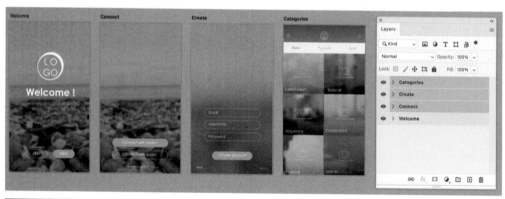

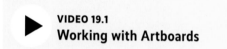

FIGURE 19.5 Exporting three selected artboards to separate JPG files

Proofing Colors Onscreen

You can use color profiles to preview how your document will look when it is output under a specific set of conditions, most commonly for print. This is called soft-proofing, and the accuracy of it depends on several factors, including your display device, the lighting conditions in your environment, and the color profiles you use. At minimum, you need to know the settings your print service provider recommends. Once you have these, you can proceed with the following steps.

To soft proof a document:

1. Choose View > Proof Setup > Custom to open the Customize Proof Setup dialog.

2. If you have a proof settings file (PSF) from your print service provider, click Load and select the file. Otherwise, set the Proof Conditions manually, specifying settings that match the output conditions you want to simulate:

   ▸ **Device to Simulate** is the output profile recommended by your print service provider. It takes into account the printing device, inks, and paper.

   ▸ **Rendering Intent** is the method Photoshop uses to convert colors from one profile to another.

   ▸ **Black Point Compensation** maps the darkest point of the source profile (working or embedded) to the output profile. Leave it on unless your print service provider recommends not to.

3. (Optional) Turn on Simulate Paper Color and Simulate Black Ink if your print service provider recommends it. Your image may look noticeably less vibrant with these options turned on, but that's actually a good thing because it sets your expectations realistically for the fact that printing processes generally cannot reproduce the full range of colors and tones that a screen can display.

> **TIP** If you have already configured your color settings so the Working CMYK Space matches the output conditions you want to simulate, simply choose View > Proof Setup > Working CMYK. Or, press Ctrl/Command+Y.

> **TIP** When Proof Colors is turned on, a checkmark appears next to the Proof Colors command in the View menu and the name of the proof preset (or color profile) being used appears next to the document name at the top of the window (**FIGURE 19.6**).

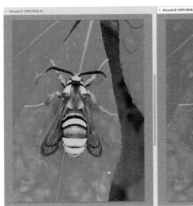

FIGURE 19.6 Left: An RGB image with Proof Colors turned off. Right: The same image with Proof Colors turned on. The CMYK profile used is shown in the document tab (or title bar in a floating window).

Preparing a File for Commercial Print

Don't assume you have to convert to CMYK for print. In many professional print workflows, RGB images are preferred, because they can yield the best results. Leaving images as RGB enables a print service provider to perform color conversions optimized for their equipment and supplies. If you *are* required to convert images to CMYK do the conversion after all other image edits and make a backup of your file in RGB mode. Converting to CMYK is a destructive change. CMYK has a smaller gamut of reproducible colors, so vibrant RGB reds, greens, and blues are compressed into a narrower (and duller) range, and you can't get them back by converting back to RGB.

To convert an image to CMYK for print:

1. Choose Edit > Convert To Profile.

2. In the dialog, under Destination Space, choose the CMYK output profile that your print service provider recommends.

3. In Conversion Options, choose any other recommended settings (**FIGURE 19.7**). If you printer did not give you recommendations, choose Adobe (ACE) for Engine and Relative Colorimetric for Intent, and confirm Use Black Point Compensation, Use Dither, and Flatten Image To Preserve Appearance are all on.

4. Click OK.

5. (Optional) If you know the characteristics of the device your print service provider will use, adjust Levels, Curves, or Hue/Saturation to maximize contrast while preserving detail in shadows and highlights.

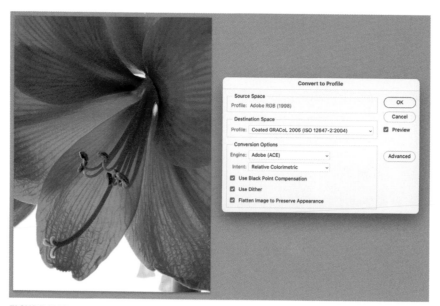

FIGURE 19.7 Instead of choosing Image > Mode CMYK Color (which doesn't give you any control over the conversion), it's better to use the Convert To Profile feature when converting an image from RGB to CYMK for professional print. Convert To Profile gives you a dialog where you can confirm and apply the settings recommended by your print service provider.

Printing to a Desktop Printer

Printing to a local desktop printer from Photoshop is like printing from any other application. Use the Photoshop Print Settings dialog to preview the printout and set options.

To print to a desktop printer:

1. Choose File > Print, or press Ctrl/Command+P to open the Photoshop Print Settings dialog.

2. In the Printer Setup area, select the printer, number of copies, and orientation. You can also access your printer's device settings here, but do not apply the same type of settings in both dialogs to avoid unexpected results.

3. In the Color Management area, you can choose whether Photoshop or the printer will manage the color conversion. Our recommendation is to use Printer Manages Colors unless you can select a profile for your printer's paper/ink combination, and the printer driver lets you disable color management.

4. Use the Position and Size options and the preview area to arrange the image on the paper. You can drag on the preview to change the image position.

5. (Optional) In the Printing Marks area, add crop marks, registration marks, a description (from the image metadata or enter one by clicking Edit), and a label (the file name as shown in the document title bar) as desired (**FIGURE 19.8**). If you're adding marks beyond the canvas, be sure the selected paper size has enough room to fit them.

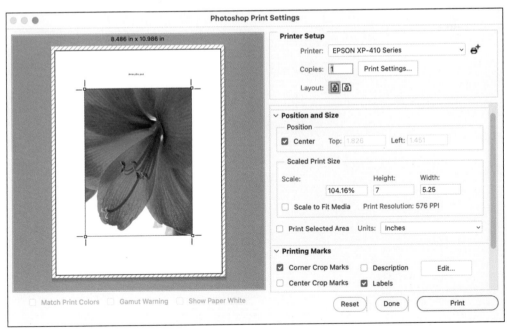

FIGURE 19.8 Use the controls in the Photoshop Print Settings dialog to set the position and size of the image on the paper, as well as add crop marks and labels if desired.

6. (Optional) In the Functions area, invert the image colors by choosing Negative. Flip the image horizontally by choosing Emulsion Down. You can also apply a background color, a solid line border, and bleed settings here.

TIP The Photoshop Print Settings dialog offers additional options, such as printing separations from a CMYK file, when you are printing to a PostScript device.

TIP Hover your pointer over each Color Management option to read a detailed description of it. Note that you may need to scroll or expand the dialog to make the description fully visible.

TIP To print just a portion of the image, choose Print Selected Area and use the cropping controls on the preview.

TIP If you know you want to use the same settings as the last time you printed, choose File > Print One Copy and skip the dialog.

Exporting to Different Sizes and Formats

You can use Export As to export an entire document or its components (artboards, layers, or layer groups) to PNG, JPG, and GIF file formats. The Export As dialog gives you the ability to scale exported files, include metadata, and convert them to sRGB.

To export a document:

1. (Optional) Choose File > Export > Export Preferences. Under Export As Location, select either Export Assets To The Location Of The Current Document or Export Assets To The Last Location Specified.

2. To export the entire current Photoshop document, choose to File > Export > Export As. If the document contains artboards, you can export them all (or any subset) in the subsequent dialog. Or, to export specific layers, artboards, or layer groups, select them in the Layers panel, then right-click and choose Export As from the context menu.

3. In the Export As dialog, set the Image Size, Canvas Size (if you need to expand or contract it separately from the image), Metadata options, Color Space (convert to sRGB or leave colors unchanged), and File Settings.

4. (Optional) To export to multiple sizes, use the controls in the Scale All area of the dialog. Click the plus button to add additional sizes to the list. Choose a Size and a Suffix for the filename (**FIGURE 19.9**).

5. Click Export.

 **TIP** When exporting multiple layers, artboards, or layer groups, you can use different settings for each one by clicking it in the left side of the Export As dialog before choosing their options.

▶ VIDEO 19.2
Exporting to Multiple File Sizes and Formats with Export As

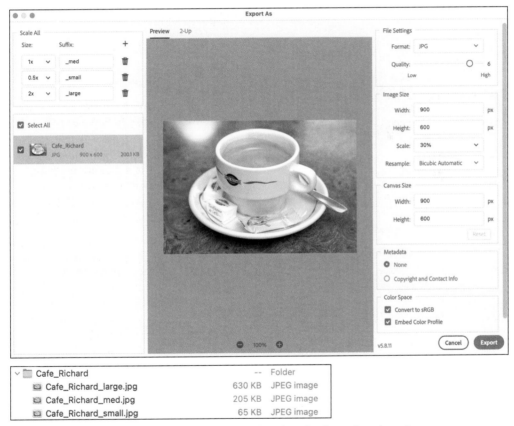

FIGURE 19.9 Use the Export As dialog to output several copies of an image in various sizes.

Using Quick Export

Quick Export can save you time if you commonly need to export files to the same location and in the same file format. You first need to set up the preferences, and then it's just a matter of choosing the command any time you want to quickly export a file. The file is immediately exported with no intervening dialog. With Quick Export, you can export documents, artboards, layers, or layer groups to PNG, JPG, or GIF formats.

To quickly export to a favorite file format:

1. Choose File > Export > Export Preferences.

2. In the Preferences dialog, choose a format, location, metadata to include, and color space to use (**FIGURE 19.10**).

3. To export the entire file, choose File > Export > Quick Export As (PNG, JPG, or GIF). Or, to export specific layers, artboards, or layer groups, select them in the Layers panel, then right-click and choose Quick Export As (PNG, JPG, or GIF) from the context menu.

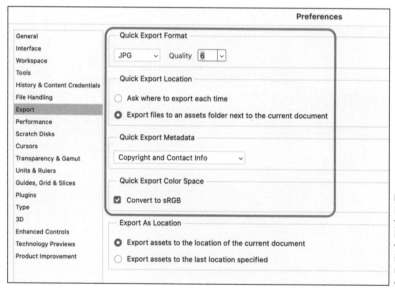

FIGURE 19.10 For Quick Export, you set up all your desired options in Preferences. That way, you can export images instantly without needing to use a dialog each time.

Exporting Layers as Files

You can export layers as separate files in formats such as PSD, JPEG, PNG, PDF, and TIFF. Files are exported in the same order in which they appear in the Layers panel from top to bottom and named with the layer names plus a prefix that you can define.

To export layers as files:

1. Choose File > Export > Export Layers To Files to open the Export Layers To Files dialog (**FIGURE 19.11**).

2. In the dialog, click Browse to choose a destination for the exported files. By default, files are exported to the same location as the source file.

▸ (Optional) Enter a name in the File Name Prefix to be applied to the exported files. Or, leave this field empty to have the filenames prefixed with an underscore and four-digit number.

▸ (Optional) Select Visible Layers Only, if desired.

▸ From the File Type menu, choose a file format and set options as necessary.

▸ Select Include ICC Profile to embed the working space profile in exported files.

3. Click Run. A confirmation dialog appears when the files are finished exporting.

TIP If you use Layer Comps, you can also export them to separate files, by choosing File > Export > Layer Comps To Files.

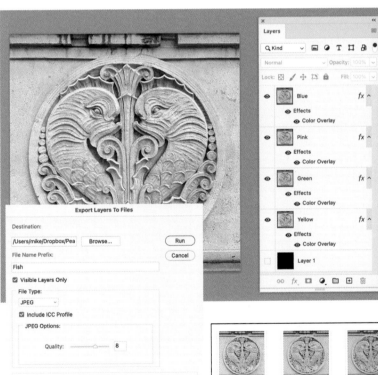

FIGURE 19.11
You can export separate files for layers in a document to various file formats and apply a prefix to the output file names.

Creating Animated GIFs

You can create a looping animated GIF from a layered Photoshop file or a video. A common scenario is to combine a series of still images taken with a mobile phone's burst feature in one layered file and then export it as a GIF.

To create an animated GIF from still images:

1. Choose File > Scripts > Load Files Into Stack. In the Load Layers dialog, click Browse to select the files you want to use. Click Open and then OK. A new document is created with each imported file as a separate layer (**FIGURE 19.12**).

2. Choose Window > Timeline to open the Timeline panel.

   ▸ Click the arrow on the button in the middle of the panel and choose

Create Frame Animation. Click the button to create a new frame animation. Then, from the Timeline panel menu, choose Make Frames From Layers (**FIGURE 19.13**).

▸ Press the spacebar to preview the animation (or click the Play button in the Timeline panel). Press the spacebar again to pause the preview.

▸ (Optional) To control the playback speed of your GIF, select all the frames in the Timeline panel, click the Frame Delay Time shown under any thumbnail, and choose a different delay time from the menu.

▸ Use the Repeat menu in the Timeline panel to select how many times to the GIF will play: Once, 3 Times, or Forever.

▸ Choose File > Export > Save For Web (Legacy).

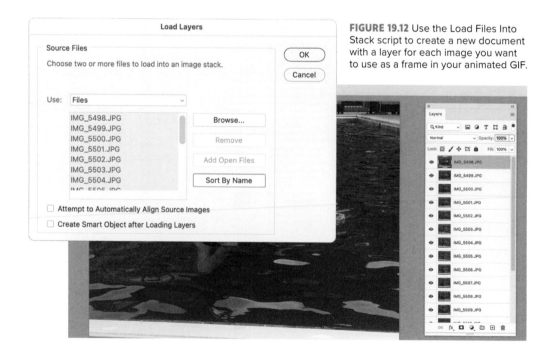

FIGURE 19.12 Use the Load Files Into Stack script to create a new document with a layer for each image you want to use as a frame in your animated GIF.

3. In the dialog, select the following options (**FIGURE 19.14**):

   ▸ From the Preset menu, select GIF 128 Dithered. From the Colors menu, select 256. From the Looping Options menu, select Forever.

4. (Optional) Use the Width and Height fields in the Image Size options to resize the file and reduce file size.

5. (Optional) To preview your GIF in a web browser, click the Preview button in the lower-left corner of the dialog.

6. Click Save.

To create an animated GIF from a video:

1. Choose File > Import > Video Frames To Layers. In the Import Video To Layers dialog, use the controls to select the portion of the video you want to use and click OK. A new document is created with each selected video frame as a separate layer.

2. Choose Window > Timeline to open the Timeline panel. The layers will automatically appear as frames in the timeline.

FIGURE 19.13 After you choose Make Frames From Layers from the Timeline panel menu, each layer is represented in the timeline.

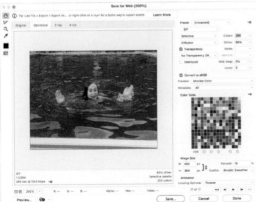

FIGURE 19.14 Use the Save For Web (Legacy) dialog to export your animated GIF.

3. Press the spacebar to preview the animation (or click the Play button in the Timeline panel). Press the spacebar again to pause the preview.

4. Use the Repeat menu in the Timeline panel to select how many times to the GIF will play: Once, 3 Times, or Forever.

5. Choose File > Export > Save For Web (Legacy), set options as for creating a GIF from still images, and click Save.

TIP To control the playback speed of your GIF, select all the frames in the Timeline panel, click the Frame Delay Time shown under any thumbnail, and choose a different delay time from the menu.

 VIDEO 19.3
Creating Animated GIFs

Using the Package Command

With the Package command you can gather all linked image assets associated with a Photoshop file in a folder along with a copy of the file to share them with a colleague or create a backup or archive. This includes linked items from Creative Cloud libraries and linked Smart Objects. (Fonts cannot be packaged.)

To package a Photoshop file:

1. Choose File > Package.

2. In the dialog, navigate to the desired location for the package and click Choose. A folder is created containing copies of the Photoshop file and the linked image assets. Items that were linked from Creative Cloud libraries are converted to linked Smart Objects with the links pointing to the packaged asset files (**FIGURE 19.15**).

FIGURE 19.15 The Package feature creates a folder named for the document, containing a copy of the document and all linked image assets, ready for sharing or archiving.

Index